KILLERS IN EDEN

To Mike

KILLERS IN EDEN

*The true story of killer whales and
their remarkable partnership
with the whalers of Twofold Bay*

DANIELLE CLODE

Danielle Clode

ALLEN&UNWIN

First published in 2002

Allen & Unwin
83 Alexander Street
Crows Nest NSW 2065
Australia
Phone: (61 2) 8425 0100
Fax: (61 2) 9906 2218
Email: info@allenandunwin.com
Web: www.allenandunwin.com

National Library of Australia
Cataloguing-in-Publication entry:

Clode, Danielle.
Killers in Eden : The true story of killer whales and their remarkable partnership with the whalers of Twofold Bay

Bibliography.
Includes index.
ISBN 1 86508 652 5.

1.Killer whale – New South Wales – Twofold Bay. 2. Whaling – New South Wales – Twofold Bay – History. I. Title.

639.28099447
Typeset by Midland Typesetters, Maryborough, Victoria
Printed by Griffin Press

10 9 8 7 6 5 4 3 2 1

ACKNOWLEDGMENTS

Without the years of work by countless researchers around the world, I could never have attempted to reconstruct the lives of the killer whales of Eden. I am indebted to all those biologists, historians, journalists, museum curators, geographers and anthropologists whose work I have drawn upon and hope that my extrapolations and interpretations have done their research justice.

A number of individuals have greatly assisted me in completing this book. It was thanks to the tireless support of Jenny Darling and Jacinta de Mace that I originally met my commissioning editor, Ian Bowring, whose enthusiasm for the killer whales of Eden rekindled my own interest. Australian Geographic generously funded my research while my colleges at the Zoology Department of the University of Melbourne kept me on the academic straight and narrow. I am particularly thankful for the support, advice and hospitality of René Davidson, whose interest in this story provided a personal insight into the Davidson family history and the legends surrounding their involvement with the killer whales. Robyn Kesby alerted me to Sue Wesson's research into Aboriginal whalers, and I am grateful to Sue for generously discussing her work with me and suggesting the probable namesakes for the killer whales. George Macdonald provided a whirlwind introduction to Haida Indian killer whale culture and animistic spiritual belief systems. Greg McKee added to my store of accumulated pictures, facts and stories about the killer whales. Ingrid Visser, John Ford and Joan Dixon corrected some of my misconceptions about killer whales, their evolution and behaviour. Emma Cotter and Narelle Segecic have done an impressive job of ironing out repetitions, ambiguities and errors in the manuscript.

Thank you all for helping this book come to fruition.

Picture credits:
Cover: 'Twofold Bay' by O.W. Brierly; killer whale
Chapter opening sketch: by O.W. Brierly from his *Diaries at Twofold Bay and Sydney* (held at the State Library of New South Wales, MLA: frame A534)
Pictures from René Davidson's *Whalemen of Twofold Bay*: cover and pages ii-iii, 20, 21, 23, 24, 25, 26, 29, 31, 32, 33, 65, 68, 104 and 154.
Illustrations/map by Ian Faulkner: pages 6, 15, 43, 50 and 105

Note to Aboriginal readers:
This book contains photographs of deceased Aboriginal men who were involved in the whaling industry of Twofold Bay.

CONTENTS

PROLOGUE

Let us enter, in literary conspiracy, the lapidary shop of English Fine Phrases. We will burgle the strong rooms of Language and ransack the drawers of Diction to find fit jewels of description. We will take sapphires and arrange their dark blue radiance in an undulated plain—we will call it the Sea. Over this we will arch a mass of turquoise, and call that the Sky. On the left we will lay down emeralds in all their hues of greenness, which will make the Land. Hang an immense brilliant [sic] in the centre of this imaginary jewel picture for the Sun; spread gold freely for the Beaches; throw in opals, chalcedony, topaz, agate, carnelian, sard, amethyst, pearls, amber, rubies, moonstones, crystals, porphyry, alabaster, marble and ebony. Scatter them freely—and you get an idea of Twofold Bay on a clear Spring morning.

E. J. Brady, *The Law of the Tongue*, p. 37

I never saw a killer whale in Eden. Not as we sailed into the idyllically calm and sheltered coves of Twofold Bay. Not during the days we stayed in the harbour, exploring the hilly streets of Eden, seeing the local sights and sailing across to the whaling station and Boydtown. Not even as we regretfully left the calm harbour. I stared and stared across the silvery waters, waiting for a tall black fin to slice through the oily surface—but nothing appeared.

It was a bright and overcast day when we sailed into Eden. The swell was smooth and unbroken and even from high up in the rigging the reflection of the clouds left an impenetrable glaze on the surface of the water. I sat on the slats of the crow's nest, my feet twined in the ratlines, and tried to pretend there might be a whale shooting along just below the surface, safely out of sight. But I knew the killer whales had gone. The majestic black-and-white predators, which had once made this stretch of the east Australian coast their home, had long since disappeared.

Ever since we'd left our home town of Port Lincoln, in South Australia, I'd been on the lookout for whales. Any whales, even dolphins, were great fun although we saw the latter with regularity. Sailing along the coast, pods of these friendly creatures would zoom up to escort us through their ranges, scooting effortlessly under the bow of the boat. I'd sit on the chains under the bowsprit, next to the dolphin-striker (which never did, fortunately), just inches from the dolphins below. Often the dolphins spun on their sides, their small eyes scrutinising this strange terrestrial creature sitting so close above them. Suddenly, without warning, the entire pod dived away into the depths as if unwilling to cross some invisible boundary. We were left to forge on through the swells alone.

Eden transformed this vague interest in whales into a fascination. Eden had a kind of mythical status in my mind, long before I ever went there. A few years earlier, when I was about ten, my father had gone to Sydney to help sail a boat around to Adelaide. He'd sent me a postcard from Eden. The tinted waters of Twofold Bay, strange ruins and rusty bits of deserted whaling stations were fascinating and enticing: images forever entwined with my delight at receiving my first letter. By that one small, random act of paternal care, Eden had acquired a magical veneer. I was ready to be impressed.

When my parents sold up everything in the late 1970s to live on a boat, people often wanted to know where we were going. Since the prevailing south-westerly winds best suit sailing east from Port Lincoln, 'Queensland' became our vague destination. Queensland is the destination of choice for many of Australia's retired, unemployed or alternative life-stylers. The northern tropical warmth is a compelling temptation in the middle of a southern winter and for people on boats it has the added advantage of the Great Barrier Reef. Not only is the reef an attraction in its own right, but it cuts out the swell and provides little scope for the generally mild winds to create large waves. Small wonder so many regard Queensland as a sailor's paradise.

The south coast, however, requires a far more rugged sailor. The entire southern Australian coast is characterised by rough cliffs—from the bared limestone teeth of the Great Australian Bight to Victoria's famous 'Shipwreck Coast', whose eroded pinnacles stand testament to oceanic fury. This coastline is exposed to the full fury of the Southern Ocean. Massive swells build across the empty nautical miles of icy water—formed in the roaring forties—rolling in effortlessly to smash against the gouged cliff faces. The open waters of the southern coast are bad enough, but the waters of Bass Strait, apparently protected by Tasmania are far worse. When large ocean swells are forced into the narrow, shallow waters of Bass Strait they turn very nasty. The shoaling waters force the swells up into short, sharp, close-packed crests, while the now submerged mountain ranges which once stretched from the mainland to Tasmania erupt as jagged islands and rocks along the eastern and western boundaries. World-weary ocean sailors describe Bass Strait as the worst patch of water in the world.

Our entire journey along the south coast, like that of so many other sailors, was a series of short dashes, from one harbour to the next, trying to outrun the fronts as they swept through every few days. We entered Bass Strait on a calm and mild autumn day. Could we get through Bass Strait without the mandatory gale? No such luck. A steady northwesterly wind saw us thundering along under a reduced rig. During the night, the strengthening wind shifted to the south-west and swept us towards the treacherous Victorian coast long before the breaking daylight we needed to negotiate rocky outcrops. My parents spent the night battling against a relentless gale determined to force us shoreward. Unable to help them, I spent the night lying on the floor with my wild-eyed and sharp-eared cat clinging to the carpet next to me, watching the bulkheads turn from white to green on every roll.

Stomach-churning experiences like this made us long to 'turn north' and escape the dreaded southern coast. Fortunately, only a few days' sailing remained to reach Gabo Island just off the southeastern tip of mainland Australia. Gabo Island marks the magical turning point at which we stopped sailing east and began to head north—away from the cold, rough, southern waters and up the eastern coast with its safe harbours, warm weather, gentle winds and calm seas. We were through the worst bit of ocean in the world. Ahead lay plain sailing, delightful seaside towns and, ultimately, the tropical delights of the Great Barrier Reef. With a sense of great celebration and anticipation, we headed north—to Eden.

Eden has a lovely harbour. Spacious Twofold Bay provides easy access to the small and aptly named Snug Cove in its heart. From the sea, Boyd's Tower stands guard on the southern promontory of the bay, overshadowed by the dark mountain flanks in the distance. Deep in the centre of the bay a little hilly point juts out into the bay, defining the beginnings of the small town of Eden which stretches up and behind it. This point shelters Snug Cove and contains the historic centre of Eden. The town is surrounded by dense forests to the west and sea to the east, both of which have provided its life blood since European settlement.

Eden bears all the hallmarks of a town which boomed late in life, and even then only in a small way. It has never been rich and has none of the fine monuments and public buildings that characterise once-wealthy 'gold' towns. Eden's historic buildings were built a bit at a time, something from here and something from there, a new veranda now and a new storey then. Few of these buildings survived the forestry-driven expansion in the 1970s which saw dilapidated shops and houses replaced by tiled shopping arcades, now home to laundromats and quiet real estate agencies. Only a few heritage buildings survive—the Great Southern Inn, a church or two, a house now run as a bed-and-breakfast. Most of Eden's architectural record is confined to small plaques noting the place where a building of interest once stood. Even the harbour in Snug Cove has been repeatedly reshaped over the years to accommodate its fishing fleet.

Our first stop upon arriving in Eden was the local post office to collect our mail. Most of the mail was for me—from my teachers at the South Australian Correspondence School. Every fortnight I sent off completed assignments for each of seven subjects. Days in harbour had to accommodate taping German classes, completing essays and struggling with maths assignments. One subject I greatly enjoyed was geography. My teacher, Mr Williams—no doubt weary of reading yet another essay on the wheat belt or cattle industry—had suggested that I substitute them with projects on local history and geography. Museums and sites of local significance were a high priority. Eden's renowned Killer Whale Museum was first on the list.

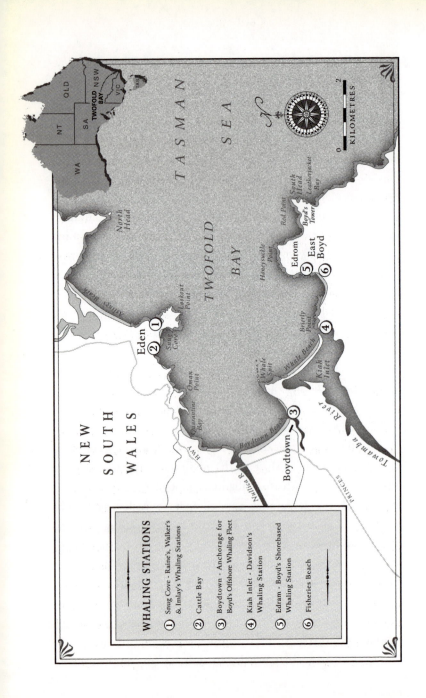

WHALING STATIONS

① Snug Cove - Raine's, Walker's & Imlay's Whaling Stations

② Cattle Bay

③ Boydtown - Anchorage for Boyd's Offshore Whaling Fleet

④ Kiah Inlet - Davidson's Whaling Station

⑤ Edram - Boyd's Shorebased Whaling Station

⑥ Fisheries Beach

NEW SOUTH WALES

TASMAN SEA

TWOFOLD BAY

North Head

Lookout Point

Eden

② ①

Snug Cove

Oman Point

Quarantine Bay

Whale Spit

Whale Beach

Boydtown

③

Boydtown Beach

PRINCES HWY

PRINCES & MONARO HWY

Towamba River

Kiah Inlet

④

Briefly Point

Honeysuckle Point

Edrom

East Boyd

⑤ ⑥

Boyd's Tower

Red Point

South Head

Leatherjacket Bay

Asling's Beach

KILOMETRES

0   2

TWOFOLD BAY

WA

NT

SA

QLD

NSW

VIC

A charming elderly man by the name of Bert Egan greeted us at the small building. The centrepiece of the museum was a massive killer-whale skeleton, about 7 metres long. 'Old Tom's' impressive set of peg-like teeth grinned an alarming welcome. His teeth, according to Mr Egan, are more than just a reminder of the whale's predatory life-style. They are also testimony to the whale's association with the whalers of Twofold Bay who, for 100 years, plied their trade with the assistance of this particular whale and its family. Tom's front and side teeth are missing and many are worn off, as if a bit had ground them down. According to local legend, Tom often took the painter of a boat, or the cable of a harpoon attached to a whale, in his teeth to pull the boats faster or help drown the whale. Such unnatural behaviour was thought to have caused Tom's worn teeth. And indeed, even the remaining intact teeth reveal a curious half-circular pattern of wear suggestive of a rope.

Bert Egan told us that as a young boy he had witnessed some of the last whale hunts in Eden. He guided us around the museum, enhancing the displays of faded photos and newspaper clippings with personal anecdotes and recollections. The story grew all the more remarkable for his telling. Every year Old Tom and his family would appear in Twofold Bay. This pod of killer whales cornered migrating baleen whales and harried them into the bay. Former whalers reported that Old Tom would often head into the whaling station at Kiah Inlet opposite Eden to alert them to their quarry's presence. According to Bert Egan, there was no mistaking the killer whale's call for attention. Lifting himself bodily out of the water, Tom would crash down on the water's surface near the whaling station in an action known as floptailing. With a cry of 'Rush-oo' the whalers would head out to the Bay with their killer whale guide alongside.

Out in the bay, the other members of the pod would be

targeting the baleen whale, harrying it from behind and underneath to prevent it diving, taking bites of its flippers and mouth
and flinging themselves on top of its blowhole to try to drown it.
When the whalers arrived, the killer whales would erupt into
frenzied excitement, leaping out the water and racing from boat
to whale like enthusiastic dogs.

Harpooning a massive 20-metre whale from a flimsy 9-metre
rowboat is a dangerous occupation. The shrimp-sieving baleen
whales may have seemed gentle in comparison to their more
aggressive cousins, but their sheer size was a weighty weapon.
Even the placid and amiable humpbacks and right whales could
smash a boat with one swipe of their powerful tails. A mean and
angry fin whale presented an even greater risk. No whaler
wanted to fall into water churning with the blood of a dying
whale and amid the gouging teeth of the predatory toothed
killer whales. But despite their legendary appetites, the killer
whales never harmed a whaler. In fact, local legend claims that
when a man went overboard, a killer whale would leave the frenzied hunting and swim gently alongside, standing guard, until a
boat retrieved them. One old whaler claimed that a killer whale
grabbed his shirt and hauled him to the surface as he sank below
the water.

Once harpooned, the baleen whale belonged to the killer
whales for the time being. The museum displays described how
the killers took the harpoon cables in their teeth, to drag the
whale deep in the Bay before feasting on its tongue. Decomposi-
tion eventually refloated the bloated carcass which would be
retrieved by the whalers and flensed for oil.

The killer whales became legendary in Eden and stories
spread of their prowess. I later heard of whaling masters who
claimed them as their 'dogs' and read that the killers would only
work for long-time local whaler, George Davidson, even taking

the painter of his boat in their teeth and towing him at speed to ensure he reached the whale first in a race with a rival company. When the killers became entangled in the harpoon ropes, newspapers reported that they lay still and calm to be disentangled by their human associates, emitting a gentle cat-like purr when freed. Aboriginal whalers believed they were the reincarnated spirits of dead whalers and named them accordingly. Tom and Hooky became local institutions for the 70 or 80 years they lived in the area.

At the time, I took such stories as gospel and reported them as such to my delighted teachers. It was many years before I returned to Eden and the legends of the killer whales, this time as a biologist with a particular interest in animal behaviour. In the meantime every time I met a whale expert, at every conference and university I attended, I would quiz them for their knowledge of this unusual and remarkable association. Responses ranged from knowledgeable to incredulous, from fascinated to sceptical. Questions had also emerged in my own mind. How could Tom have lived for 70 to 80 years when male killer whales only live for 30 or 40? Could he have been an old female (which lives twice as long as her mates)? Were the whales really cooperating or just exploiting the whalers? Or were the whalers just exploiting natural killer-whale behaviour? What was the relationship between the killer whales and the local Aboriginal people? Did this human–animal hunting association pre-date the Europeans who documented it and claimed it as their own?

Historical events are difficult for scientists to study. Science demands replication, control, observation and corroboration. A unique association between two groups of different species over a century ago is problematic to address scientifically. Whales themselves are among the most challenging of mammals to study, and for many years virtually nothing was known about killer whales.

But times have changed and in the last twenty years killer-whale research has expanded dramatically. In British Columbia, hundreds of killer whales have been studied individually since the 1970s, becoming the best-studied population of killer whales in the world. Their families, life histories, interactions and communications have all been the subject of intense and fruitful investigation. Other populations have also received several years of study. French scientists have documented the remarkable hunting strategies of killer whales off the Crozet Islands in the southern Indian Ocean. Killer whales off the coast of Norway, Argentina, California and even Antarctica are starting to receive similar attention. Modern methods of genetic analysis promise to reveal information about evolution, patterns of mate choice, paternity and family histories, which could previously only be uncovered after decades of painstaking field-work. The more we know about these impressive creatures, the more there is to know and the more I am impressed by killer whales.

There is now a relatively good understanding of some aspects of killer-whale society, behaviour and physiology, at least from some parts of the globe. The killer whales themselves may have disappeared from Eden, but the behaviour of their relatives in other oceans gives us an insight into how and why this seemingly remarkable association between human and whale developed. In this case at least science can offer a new view of history, not from the perspective of incredulous reporters, battle-weary whalers or entranced spectators, but more from the perspective of the whales themselves.

This is the story of Tom, Hooky, Humpy, Cooper, Typee, Jackson, Stranger, Big Ben, Young Ben, Jimmy, Kinscher,

Sharkey, Charlie Adgery, Brierly, Albert, Youngster, Walker, Big Jack, Little Jack, Skinner and Montague. This is the story of all the killer whales of Eden, those known and those whose names have been lost; their ancestors and perhaps their descendants; their families, their histories and their natures; as best I can reconstruct them over the intervening decades.

Hamlet:	*Do you see yonder cloud that's almost in shape of a camel?*
Polonius:	*By the mass, and 'tis like a camel, indeed.*
Hamlet:	*Methinks it is like a weasel.*
Polonius:	*It is backed like a weasel.*
Hamlet:	*Or like a whale?*
Polonius:	*Very like a whale.*

William Shakespeare, *Hamlet*, Act III, Scene ii

1

THE FAMILY ALBUM

Two more fins appeared between the two boats—one sharp-pointed with a little round knob and the other just an ordinary fin dropping sideways.
    'We've got company,' George remarked, 'Tom and Hooky.'
                                    Tom Mead, *Killers of Eden*, p. 71

The century of whaling in Eden encompasses two, even three, killer-whale generations. None of the adult killer whales witnessing the arrival of European whalers in Twofold Bay in 1828 would have seen its conclusion in 1926. Successive generations of whales must have been recruited into the hunting party, learning to cooperate with the whalers from the older whales in the pod, who in turn had learnt to trust and work with the whalers of Twofold Bay.

Others, like Tom Mead and Alex McKenzie, have documented the human history of whaling in Eden far better than I can, but I was surprised to find that no-one had ever brought together all the information on the individual killer whales which made whaling in Eden unique. Trawling through the literature to pull together names, anecdotes and descriptions of as many whales as possible became my first goal in attempting to decipher the hidden history of Eden's killer whales.

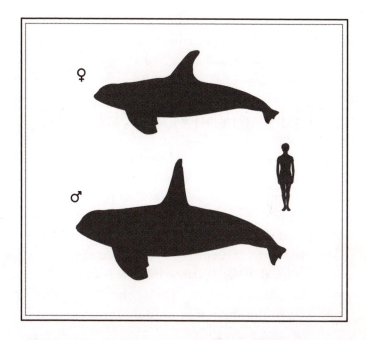

Killer whales are fairly easy to distinguish from other whales. While most whales have fairly small, streamlined dorsal fins, the massive triangular fin of a male killer whale stands taller than most people. Even female killer whales have fins up to 1 metre high. Their striking black-and-white markings are also highly distinctive, particularly when spyhopping with their heads out of water, or floptailing and flinging themselves skyward to reveal their white abdomens.

These features not only make it easy to recognise the killer whale species, but with practice, also enable individual killer whales to be identified. Biologists today use photos of dorsal fins to identify the animals they study. A century ago, the whalers of Eden used the same characteristic to recognise and name regular members of the pod and observe individual patterns of behaviour. Some of the killers led the pod or smaller hunting groups.

Some performed particular tasks during the hunt, while others, like Tom, showed flamboyant (and at times annoying) personalities. Without their characteristic dorsal fins, such individual differences would have been difficult to discern and the relationship between the whalers and the killer whales might have been far less personal than the recollections of, and documents from, the time suggest it was.

Some killer whales, such as Humpy, may have been named after the physical characteristics of their fins while others, like Youngster, may have been named for their position within the pod. But most were named after deceased Aboriginal whalers who provided the bulk of the labour for whaling in Eden from the first days in the 1830s to its cessation in the early 1900s.

Oswald Brierly, whaling station manager from 1843 to 1846, was the first European to notice how Aboriginal Australians identified the killer whales as deceased ancestors:

> The natives of Twofold Bay regard the killers as incarnate spirits of their departed ancestors, and in this belief, they go so far as to particularize and identify certain individual killers. (O. Brierly, 'Reminiscences of the sea: About Whales' MLA546: frame 10)

This pattern of nomenclature ensures that a killer whale is unlikely to have reached maturity (and attained its distinctive identifiable fin) before the death of its patronym. Identifying these namesakes, their lifespans or date of death where known provides an indication of when an individual killer whale matured within the pod. This difficult task has been completed for many whales by cultural geographer, Sue Wesson, whose report on the Aboriginal whalers of Twofold Bay for the NSW National Parks and Wildlife Service is the first thorough account of the important contribution of Aboriginal Australians to whaling in Eden.

Of the 25 to 30 killer whales known to visit Eden around the end of the nineteenth century we know the names of 21. Of these animals, we have photos of six and anecdotes and observations of about twelve; the rest are known only by their names. Information on the individual killer whales in the pod is sketchy, with the exception of Tom. Nonetheless, coupled with knowledge of patterns of family life in other killer-whale populations it provides some clues about the probable structure of the Eden pod, at least in its later years.

The earliest record of named individual whales is in an article by H. S. Hawkins and R. H. Cook from 1908 published in the monthly *Lone Hand* magazine. The article, based on discussions with George Davidson and his whalers, named six whales—Walker, Big Jack, Little Jack, Skinner, Hooky and Humpy—while noting that about 30 animals in total visited Twofold Bay each year and had been known individually for over 25 years. Another contemporary account of the Eden killer whales by E. J. Brady appeared the following year in *Australia Today*. It recorded the pod size as 'over twenty' and described a number of individuals including five not mentioned in the previous article. 'Hookey' and 'Old Humpy' were mentioned again, as well as Stranger, Cooper, 'Kinchen', Tom and Jackson.

In the early 1900s, photographers including Charles Wellings and W. T. Hall accompanied and observed the Davidson whalers on a number of hunts in Twofold Bay, providing the first photographic evidence of killer-whale activity. Wellings also apparently made a movie of the Eden killer whales during a hunt, from which a number of stills survive. The film was screened privately a number of times, but regrettably no extant copy of the movie has been located, creating the tantalising yet unfulfilled prospect of visual evidence of the killer whale's activities. The photos taken in the early 1900s nevertheless provide one of the few physical representations of the killer whales and allow us to identify many of the individual animals, particularly Tom.

The only other accounts to record details of individuals were published after the demise of whaling at Eden and hence rely on the recollections of the author and/or local whalers. David Stead described fifteen named individuals in his 1933 book on sea life—several pages of which were devoted to the Eden killer whales. Big Ben, Typee, Jimmy, Charlie Adgery and Montague were here mentioned for the first time, along with those previously known. Slight variations in name (e.g. 'Hookie' and 'Kinchie' or 'the Kinchen') suggest that Stead probably acquired these names from local whalers rather than just using the previously published lists. Much of the information in Stead's account may well have come from whaler George Davidson, who died in 1952.

Similarly, a retrospective account of the Eden killer-whale story by Charles Wellings, published in 1944, was based on information he gained while staying with the Davidsons and participating in a number of whale hunts from 1910 to 1913. His list of names added two more animals to the list (Albert and Young Ben). Anecdotal reports compiled by the Eden Killer Whale Museum add Youngster, Sharkey and Brierly to the list, bringing the total number of names to 21. Given that the entire pod around the turn of the century consisted of no more than 30 animals, these names may well constitute all the adult killer whales who actively participated in the hunts.

EDEN POD TIMELINE

Date	Records
1840	Hundreds of killer whales present
1842–4	Lengthy descriptions of killer whales assisting hunts by Brierly
1878	27 killer whales present in three pods, each led by Hooky, Stranger and Cooper Tom noted as present
1880s	Towards the end of this decade, 20–30 killer whales present, including Humpy, Stranger (pod leader), Jackson, Kinscher, Tom, Cooper and Hooky, Walker, Big Jack, Little Jack, Skinner (published 1908–09 but probably refers to earlier time)
1895	Cooperation of killers mentioned in local paper
1901	Typee stranded with grampus and killed
1902	Only about seven animals return this year, including Tom, Humpy, Hooky, Charlie Adgery, Young Ben and Kinscher
1907	Stranger killed by Botany Bay fisherman in August
1909	Reference to Humpy freed from line by George Davidson
1910–12	Movie shot of killer whales by C. E. Wellings
1912	Killer whales stop appearing regularly, only a few seen occasionally
1919	Tom snatchs harpoon line in his mouth and pulls it out of boat
1923–8	Only three whales return, including Tom
1928	Only Tom and Hooky return
1929	Only Tom returns
1930	17 September Tom found dead

Tom
Present at Eden from 1860s to 15 September 1930

A WHALE KILLER
"TOM" AT EDEN, N.S.W.

Tom has been immortalised in Eden folklore. His skeleton represents the most direct link with this remarkable part of Eden's past. Although he was reputedly seen in the area for 80 or 90 years, he probably had a much shorter life of about 60 years at the most. But Tom's longevity is not the only reason for his notoriety. During his lifetime he was renowned as the 'prankster'—a killer whale whose apparent sense of humour amused and exasperated his human companions. Whether taking unsuspecting boats on a wild ride, protecting whalers from sharks, or 'floptailing' in Kiah Inlet to call the whalers out to the hunt, Eden's affection for Tom was apparently reciprocated. While his killer-whale companions deserted Eden for better hunting grounds, Tom returned every year until his death.

Hooky
Present in Eden from 1870s to 1926

Hooky (also Hookie or Hookey) was first identified in the published literature in 1908, shortly before this photo was taken around 1910 by Charles Wellings. Hooky, a skilled whale-hunter, was credited with having saved a drowning whaler (see Prologue), and was one of the best known Eden killer whales.

Hooky's tall dorsal fin was bent forward at almost 45 degrees and to the right. The height of this fin suggests Hooky may have been a male, although it is difficult to be certain from the angle of this photograph. His distinctively damaged fin made Hooky easy to recognise but it also increased the likelihood of two killer whales being known as Hooky over the course of Eden's whaling history. Dorsal fin injuries such as Hooky's appear to be relatively common among whale-hunting killers and might have caused confusion between two similarly afflicted individuals.

Hooky may have been named for his physical deformity, however Hooky was also the name of an Aboriginal whaler who was listed in the records for a distribution of blankets in Eden on 9 June 1859. If this man was the whale's namesake, Hooky probably sustained the distinctive dorsal fin injury some time after 1859 and before the 1890s by which time we might assume the Aboriginal whaler would have died. This is consistent with George Davidson's recollection to Charles Wellings that Hooky was the leader of one of the pods in 1878 when he (Davidson) first began whaling, and with contemporary accounts of Hooky in the *Lone Hand* and *Australia Today* articles of 1908 and 1909. Assuming Hooky was a mature animal when leading the pod in the late 1800s, these dates suggest Hooky lived for at least 60 years. Hooky was last seen, as an obviously aged whale, in 1926.

Hooky was Tom's final companion in Twofold Bay and their continued close association up until their last days suggests a close family relationship. They were both at least 50 to 60 years old when they died, with local legend suggesting even older ages. They may have been brothers or, if Hooky was actually a female, mother and son. Mother–son bonds are probably the strongest of all family ties in killer-whale society and Hooky's status as a pod leader (typically accorded to the oldest animal which, by virtue of their longer lifespans, tends to be a female) is also suggestive of a mature female.

Humpy
Present in Eden from 1850–80 to 1926–27

Humpy's badly damaged dorsal fin made recognition easy but confirmation of his presumed gender difficult. This picture (which appears to have been retouched) from Charles Wellings's article reveals a relatively long dorsal fin, bent almost completely over. Many locals claimed that Humpy, like Hooky and Tom, had been present for a long time, up to 60 or 70 years prior to his disappearance between 1926 and 1927. He was mentioned in accounts from 1908 to 1909 as an outstanding assailant who often led the attack; rushing in and taking great bites from the victim's tongue, lips and flippers.

Humpy once became entangled in the line attached to a whale. Rather than panicking, Humpy lay calmly alongside the whaleboat, while the whalers slackened the line and George Davidson untangled him. When freed Humpy immediately returned to the hunt and the story, as told by Davidson and his crew was related to journalist E. J. Brady. Humpy continued visiting Eden until the 1910s and may have been one of the last three killer whales regularly seen there.

Cooper
Present in Eden from 1850s to early 1900s

Cooper is first mentioned in the published literature in 1909, although George Davidson told Charles Wellings that Cooper led one of the sub-pods in 1878. This photo of Cooper, taken in the 1900s by Charles Wellings, reveals an animal with a moderately tall, straight fin—probably a male. Cooper was named after the Aboriginal whaler Bunuangi, who was born in 1807 of the Weecon tribe of Snug Cove. Bunuangi worked for whaler George Imlay, and known by the English name of Cooper. His family was involved in whaling at Eden for three generations. A census of Aboriginal whalers in 1841 noted Bunuangi's age as 34. With short average lifespans, an Aboriginal whaler like Bunuangi may have died in the mid-1800s, suggesting that Cooper reached maturity in the 1850s. Cooper was probably an old animal when this photo was taken in the early 1900s.

Typee
Present in Eden from 1844 to 1901

Typee was a male killer whale with a distinctive hook on his fin tip. There are a number of candidates for his namesake. 'Typee' might have been derived from Toby—specifically Toby the King, or Toby Blue, also known as Buginburra who was born in Snug Cove around 1822. Another Aboriginal whaler, Calamone, born in 1799 at Snug Cove, also went by the name of Toby, or Toby Fillpot. Calamone may have been the man whose death was reported on 19 August 1844 in Oswald Brierly's diary. Brierly referred to the whaler as 'Teapot', a derivation of Toby or Toby Fillpot which may well have given rise to Typee.

Typee's death after stranding while chasing a small whale into the shallows was recorded by Charles Wellings as occurring in 1901. If Typee reached maturity in 1844, after Calamone's death, he would have been a very active elderly whale of 63 at the time of his death. More likely he died in his prime—in his forties—having reached maturity at around six years of age after Buginburra's death, assumed to have occurred in the 1860s.

Although Charles Wellings identified Typee as the stranded whale, when his daughter, Mary Mitchell, reworked his article into a booklet for publication, she switched the names of the two killer whales and identified the stranded whale as Jackson—the name by which it was known in all subsequent literature. Tom Mead, who based hid dramatised history of Eden whaling on interviews with George Davidson and his family, also identified the stranded whale as Jackson, not Typee.

Jackson
Left Eden by 1912

In his original article, Charles Wellings described how Jackson fouled a buoyline but was rescued by the whaling crew, lying still until freed and then 'flop-tailing' in apparent delight at his release. (In later accounts this story is attributed to Typee).

Although this photo of Jackson reveals a moderately elongated dorsal fin, probably of a male, the ratio of the fin's height to its width at the base is low: suggesting that he may have been a young male when the photo was taken. No other information on Jackson appears in the literature and the origins of his name remain obscure.

Stranger
Present in Eden from 1850s to 1907

Stranger may well have been the matriarch of the entire Twofold Bay clan. She (or he) led one of the three sub-pods described in 1878 by George Davidson to Charles Wellings, but was also described as the leader of the entire pack in 1909. Stranger kept to the front of the baleen whale apparently leading the killer-whale pod during the hunt. Although her behaviour and longevity suggest she was female, Tom Mead described Stranger as having a long, straight fin with a diamond shaped top, typical of males. David Stead recorded that this whale was reputedly killed by a Botany Bay fisherman in 1907. The death of such a significant individual in the pod may have disrupted the behaviour of the remaining family members and contributed to the killers' desertion of Eden in the early 1900s.

Why she was called Stranger is open to speculation. Perhaps it is an unrecognizable corruption of an Aboriginal name or perhaps she matured at a time when no deaths had occurred among the Aboriginal whalers.

Big Ben
Present in Eden from mid-1800s to early 1900s

Big Ben also came to an untimely end, stranded on rocks in Leatherjacket Bay (Mowara Point). David Stead may have been describing Big Ben when he wrote that one of the Eden killer whales 'in the violence of its endeavours to drive a small whale ashore, also stranded itself and died'. Also known as Old Ben, this whale was described in Aboriginal oral history, related by Percy Mumbulla, as a 'wizard' during the whale hunts. Big Ben may have been one of the older whales in the pod at the turn of the century, if he (or she) was named after an Aboriginal whaler called Ben, whose real name was Beerhemunjie. Beerhemunjie was probably born between 1804 and 1808 at Cape Howe and was still alive in 1844. Big Ben might have taken on his name in the mid-1800s.

Young Ben
Left Eden by 1912

After Big Ben's death, Young Ben, a young male described as having a long, straight, sharp-pointed fin, disappeared. Perhaps Big Ben was Young Ben's mother, or at least a close relative. Young Ben may have returned at a later date as he is listed as one of the last seven whales to return to Twofold Bay in 1912.

Kinscher
Left Eden by 1923

Kinscher, also known as Kincher or 'the Kinchen', was described as a 'small-finned' whale and so was probably a female. Her name might derive from the colloquial expression 'kindchen', common at the time, meaning small child or baby. This name might refer to her small fin and perhaps her smaller size in comparison with some of the larger males. Kinscher was first mentioned in publications in 1909. During a hunt, Kinscher typically threw herself bodily over the victim's blowhole in an effort to impede its breathing. Kinscher was one of seven killer whales still present in 1912, however she stopped visiting Twofold Bay by 1923.

Jimmy
Present in Eden from late 1880s to early 1900s

Jimmy was one of many killer whales to foul the lines fastened to a dead whale. If this happened during the hunt, the whalers were able to untangle the killer whales and, like Humpy, they escaped unharmed. Jimmy, however, became entangled in the buoyline attached to a dead whale after the carcass had sunk and the whalers had left. The drowned killer was found when the whale refloated and was pulled to shore. It is testimony to the value placed on these killer whales that a tribute to Jimmy was carved in the window ledge of Boyd's Tower. Another window of this lookout tower bears a tribute to the only other 'whaler' recorded as losing his life in pursuit of whales at Twofold Bay, 22-year-old Norwegian Peter Lia, on 28 September 1881. (There are reports

of Aboriginal whalers being killed while whaling but their details have not been recorded).

There were so many whalers known as Jimmy that it is impossible to tell who the whale was named after. The most likely candidate was Quira, born around 1821 in Snug Cove, who went by the name of Jamie Grey. He was still alive in 1883. Two other Aboriginal whalers known as Jimmy, Neerima (pictured) and Ananzi, were both named after whaling-station owner James Imlay, and were both alive in 1859 at the ages of 44 and 52 respectively. Older still were Pazur (Cabon Jamie), born around 1799, and Panbrowin (Crockmic Jamie) born around 1804. Parin Kingalin (Jamie Nerang) was a younger man born in 1825.

Sharkey
No dates

Sharkey's name suggests she had the small curved shark-like fin characteristic of female killer whales, but no other records of her have been found.

Charlie Adgery
Present from 1905, left Eden by 1923

Charlie Adgery was probably a young whale who arrived in the pod or reached maturity in the early 1900s. He was renowned 'for the great impetuosity of his movements' D.Stead, (*Giants and Pigmies of the Deep*, p. 43). Charlie Adgery was one of seven whales still present in 1912 but which no longer returned by 1923. He was named after Charlie Adgery (pictured) who worked for the Davidsons (along with his son Charlie Adgery junior). Charlie Adgery, the whaler, was born between 1835 and 1850 and worked in Eden as a groom before becoming a whaler in 1872. During the off-season he worked as a servant. Charlie Adgery died on 8 January 1905 at Wallaga Lake north of Bermagui.

Brierly
Present from 1899, left Eden by 1912

Brierly is listed as one of the Eden killer whales in the Eden Killer Whale Museum. This young whale was probably first named in the 1900 whaling season after the death of an Aboriginal whaler named Brierly on 13 August 1899 (buried at East Boyd). This man was probably born between 1838 and 1848, during the tenure of station manager Oswald Walter Brierly (pictured) at Boydtown and was probably in his fifties when he died. He had a long involvement in whaling, working for the Davidsons at Kiah Inlet.

Albert
Present from late 1800s, left Eden by 1912

Charles Wellings listed a killer whale by the name of Albert but nothing else has been recorded about this animal. The best known whalers called Albert were two Aboriginal men—Aden Albert Thomas (born 11 February 1878) and his son Albert (born 1902)—both of whom worked for the Davidsons for many years. However, Albert Thomas senior is thought to have died in 1964 at Wallaga Lake, and although Albert Thomas junior is pictured in whaling crews from about 1915, he would still have been a young man when whaling stopped in Eden. Neither are likely candidates for the namesake of the killer whale Albert, however he may have been named after Aden Albert Thomas's father, who might also have been a whaler.

Men standing in whaleboat at try works: (L to R) Bill Thomas, C.E. Wellings, Albert Thomas Sr., Albert Thomas Jr.

Youngster
Present in Eden in 1900s

A whale called Youngster is included in the Eden Killer Whale Museum as a member of the pod although no other information has been recorded. This whale may have been a smaller female or a younger animal which matured and was named in the latter days of the pod's activities in Twofold Bay.

Walker
Present from 1900, left Eden by 1912

The Aboriginal whaler Edward Walker (also known as Moringa) was born around 1815 to 1820. He died at Wallaga Lake in 1900. The killer whale Walker was therefore probably a relatively young animal when whaling ceased in the early 1900s.

Big Jack
Present from late 1800s, left Eden by 1912

Big Jack was first named in the literature in 1908. Judging from the name, Big Jack may have been a mature male. There were many whalers called Jack who could have been namesakes for either Big Jack or Little Jack (see below). Jack Būnjil, Jacky Wyman, Jacky Barrett and Jack (Peeginyyan) were all born in the early 1800s and may have been namesakes for a mature whale like Big Jack.

Little Jack
Present from 1900s, left Eden by 1912

Little Jack might have been a smaller female or, if Big Jack was a female, Big Jack's offspring. Little Jack might have been named after later whalers such as Jack Hoskins (who died in Bega in 1900) or Jack 'Jamby' Lawson, who was born around 1860.

Skinner
Present pre-1908, left Eden by 1912

Skinner was named as a member of a 30 or more strong pod of killer whales (along with Hooky, Humpy, Big Jack, Little Jack and Walker) in 1908. Nothing else is known about this animal.

Montague
No dates

Montague is listed by David Stead as having been one of the principal whales of the hunt and well known, although he does not describe any other details about this individual.

Not all of these individuals would necessarily have been in the pod at the same time. Some animals may have come and gone, joining the pod for a few seasons before departing. Small families of killer whales may have combined with other families to hunt whales in Twofold Bay, separating again when they left the bay. But generally, the pod seems to have been fairly stable at about 30 animals for many years before declining to just seven animals in 1912, and just three by the 1920s, with the last killer whale, Tom, dying in 1930.

*A*t the time, (probably April 1952) the research vessel was under way when I noticed a great deal of splashing between our vessel and shore. It looked as if a fast moving school of large fish was heading towards shore. I put the binoculars on this 'school' and saw 10 to 15 sea lions heading towards shore as fast as I have ever seen sea lions move.

They appeared to be skipping over the surface of the water torpedo-like in low angle leaps. Hearding [sic] them were several (5 to 7) killer whales apparently enjoying themselves. It was obvious the sea lions were terrified. The killer whales appeared to stay in a crescentrix formation at the rear and sides of this closely packed group of sea lions.

Occasionally a killer whale would dive under the sea lions and come up under one. When this happened the sea lion was either bumped or thrown several feet into the air. It did not seem however, that the killer whale was attacking the sea lion, rather it appeared as if the killer whale was enjoying a game and that the frightened response of the sea lion, resulting from the contact, was stimulating to the killer whale.

The game lasted for about a mile or so until the sea lions were about a city block off shore, at that time the entire group of killer whales attacked and for a few minutes not much could be seen but flying spray and large heaving bodies. The water over several hundred feet was churned into a bloody froth, then there were no more sea lions visible.

J. H. Fitch, Supervisor of Investigation,
Californian Department of Fish and Game, quoted in
H. P. Wellings, *Shore Whaling at Twofold Bay*, pp. 9–10

2

DEMON DOLPHINS

*When the lark and the whale were men, they fought against each
other. The lark speared the whale twice in the neck. The whale,
finding itself sorely wounded, made its escape, jumped from pain
into the sea, became a whale and spouted through the two wounds
water to heal them; but in vain, till this day.*

Adelaide Aboriginal mythology reported by
Christian Teichelmann in 1840, cited in Philip Clarke,
'The Significance of whales to the Aboriginal people of
southern South Australia', p. 22

The history of the Eden killer whales started long before the
advent of European whaling in Twofold Bay. Killer whales had
probably been visiting Twofold Bay when human ancestors were
just taking their very first steps on two feet. Even before Twofold
Bay existed, killer whales had been patrolling the world's oceans.
Killer whale history began with the origin of all the dolphins
and whales, cetaceans, 60 million years ago. At this time, great
families of large grazing animals swarmed across the land,
feeding on the rich plant life. And where there were large herbivores, large carnivores soon arrived to eat them. Today, most
large herbivores are 'ungulates', typically hoofed animals like
horses and deer, while all the big meat-eating mammals belong

to a group called 'Carnivora'. But when these groups first emerged, their roles were less clearcut and many of the dominant meat-eaters were not 'carnivores' at all, but ungulates. These meat-eating terrestrial ungulates appear to be the unlikely-looking ancestors for the modern whale family.

Hoofed predators known as mesonychids dominated the northern hemisphere landmasses. But as the climate became cooler and drier and the faster, more agile Carnivora became more competitive, the mesonychids began to die out. The only mesonychid survivors were small, specialised aquatic meat-eaters, similar to modern otters, which lived in the shallow brackish estuaries of the tropical Tethys Sea (between ancient Laurasia and Gondwanaland). Here, along what is now the Mediterranean and Asian coast, they had the 'aquatic carnivore' niche all to themselves. Their greatest competitors, the large aquatic reptile predators, such as ichthyosaurs and pleisiosaurs, had mostly been wiped out with the dinosaurs during the great Cretaceous extinction.

These mesonychids are thought to be the ancestors of larger creatures adapted to aquatic living which moved along the coast into oceans to the west, east, north and south as the Tethys Sea expanded. Known as proto-whales, or archaeocetes, they probably swam with an eel-like motion. They still used all four limbs to swim even when their hind limbs became smaller. With tiny heads on huge 20-metre bodies, the archaeocetes bore little superficial resemblance to modern whales, whose large heads may comprise up to one-third their total body size. But gradually, the archaeocetes became more whale-like. Their upper jaws broadened or elongated. Their teeth changed from the various shapes specialised for cutting, slashing and grinding, to uniform rows of pegs ideal for both impaling small slippery fish and invertebrates and ripping chunks from larger prey.

WHALE EVOLUTION TIMELINE

Epoch	Million years ago	Event
Jurassic	150	Gondwana separates from other landmasses
	145	
	140	
	135	
Cretaceous	130	
	125	Australia and Antarctica begin to separate
	120	First birds appear
	115	
	110	
	105	
	100	
	95	Marsupial and placental mammals appear
	90	
	85	
	80	
	75	
	70	
Palaeocene	65	
	60	Last dinosaurs, 1st primates & mesonychyids
Eocene	55	
	50	Archaeocetes appear
	45	Australia and Antarctica separate
	40	
Oligocene	35	Modern whales appear
	30	Squaladonts appear
Miocene	25	First seals appear
	20	Hominids appear
	15	Killer whales appear
	10	
Present	5	
	Present	Homo sapiens appear about 200 000 years ago

For the next few million years, the fossil record falls silent. From the time of the Eocene to the Oligocene epochs, 38 million years ago, whale evolution is a mystery. When the fossil record recommences, it shows that the archaeocetes had almost completely disappeared and their presumed descendants, the modern whale families of Mysticeti (baleen whales) and Odontoceti (toothed whales), had appeared.

Until now cetacean evolution had largely occurred in the northern hemisphere, particularly around India, Pakistan and the ancient Tethys Sea. But the expansion of modern whale families was tied to geological and climatic change in the southern oceans, and particularly around Antarctica.

The massive southern continent of Gondwana divided the ocean's waters, forcing the cold polar waters to circulate north to be warmed near the equator. In the 150 million years following its split from Laurasia, various continents broke away from Gondwana, leaving Antarctica stranded in the Southern Ocean—Africa and South America breaking off first, then Australia which finally broke off and drifted north 45 million years ago. As the landmasses separated and the climate cooled, the great Southern Ocean began to circulate around the polar continent. No longer warmed by tropical waters forced south by Gondwana, Antarctica froze. The circum-Antarctic current drew cold-water nutrients up from the bottom and into the more northern reaches creating great upwellings of food, particularly plankton.

This abundance of planktonic food promoted the evolution of many new predatory species. Fish, squid and cuttlefish flourished on the krill. Seabirds flourished on the fish and squid, which also attracted a new group of previously terrestrial Carnivora, the seals, back into the water. As aquatic as these new arrivals soon became, they proved to be little competition for the whales which had already spent 25 million years adapting to their marine environment.

By the early Miocene epoch, the whales were quite different from their archaeocete ancestors and considerably more aquatic than their Carnivora rivals. Their nostrils had moved to the top of the head and could be sealed against the water. Their hind limbs and pelvis, which initially shrank, had disappeared altogether. Body hair had been lost and blubber gained. The whales had become more hydrodynamic, using their enlarged tail-flukes for propulsion and a dorsal fin for control. In smaller species the dorsal fin had the added advantage of regulating temperature via blood vessels on the surface. While still needing to breath air, whales had improved their capacity to stay under water with small, rigid lungs capable of staying inflated under pressure. Many specialised structures relating to diving and echolocation appeared around this time, including the fatty tissue around the snout. The characteristic 'melon head' of many modern whales had appeared. Their sense of smell and taste had regressed while hearing had become highly acute. Every aspect of their lives had become aquatic. They grew, fed and bred in the water—even giving birth at sea.

The giant baleen whales (mysticetes) developed their remarkable sieving technique to feed on the new abundance of tiny prey. A fossil *Mammalodon colliveri*, found in the sea cliffs of Torquay in Victoria, is the most primitive mysticete known. It still has the well-developed teeth of its ancestors and lacks the distinctive baleen attachment mark on its palate of the modern mysticete whale (which it otherwise resembles), although it may have had baleen between its teeth.

The earliest toothed whales were squalodonts, or short-beaked whales, with triangular shark-like teeth and a resemblance to modern killer whales. Fossils of one species of

BALEEN WHALES AND TOOTHED WHALES

Toothed whales, or odontocetes, are characterised by their peg-like teeth, which are nearly always used to hunt fish, although squid, marine mammals and other prey are also hunted by some species. Most toothed whales have rows of uniform teeth, but some species, like the narwhal and many beaked whales, have developed sexually dimorphic teeth patterns—some enlarged and some residual. All toothed whales have a single blow-hole.

Baleen whales, or mysticetes, have no teeth but possess instead hundreds of fine hairy plates attached to the upper jaw. Bristles on these plates interlock forming a fine sieve for filtering planktonic shrimp-like krill which form the primary food of baleen whales although schooling fish, crustaceans and copepods are also important to some species. Baleen whales all have paired blow-holes.

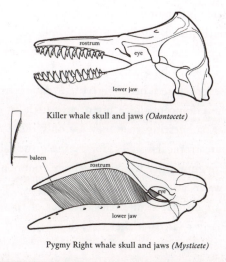

Killer whale skull and jaws *(Odontocete)*

Pygmy Right whale skull and jaws *(Mysticete)*

Prosqualodon have been found in Australia, Patagonia and New Zealand, providing us with the first known record of a species with a circum-Antarctic distribution. Unlike its northern hemisphere siblings, this species had a short, broad snout and its facial bones were pushed backwards and scooped-up like modern echolocating whales, suggesting that it probably used sonar. But despite the abundance and diversity of the squalodonts, most species disappeared by the Miocene, superseded by more successful descendants. The modern narwhals, beluga and beaked whales, as well as the smaller dolphins and porpoises, all bear the hallmarks of a squaladont ancestry. The giant sperm whales are also related to the squaladontids, but probably diverged much earlier (perhaps as much as 30 million years ago) and bear a less distinct resemblance to their smaller relatives.

In the fifteen million years since the appearance of the modern toothed whale families, many modern species have evolved. Modern dolphin, with their characteristic size and shape, proliferated. But along with these dolphin species are a few so unusual and different from the rest that some taxonomists classify them in their own family, the Globicephalidae.

The Globicephalidae differ from the other dolphins in being larger (over 2.5 metres long), having blunt heads, fewer teeth and three or more fused neck vertebrae. Apart from this, members of the Globicephalidae often have little in common with each other, suggesting that they are only quite distantly related. Most are tropical species, for example the false killer whale and the pygmy killer whale, which are commonly found in the belt of warm water that circles the globe at the equator. Despite their superficial similarities, neither of these species are actually closely related—either to each other, or to the true killer whale after which they were named.

Similarly the melon-headed whale, another tropical dolphin,

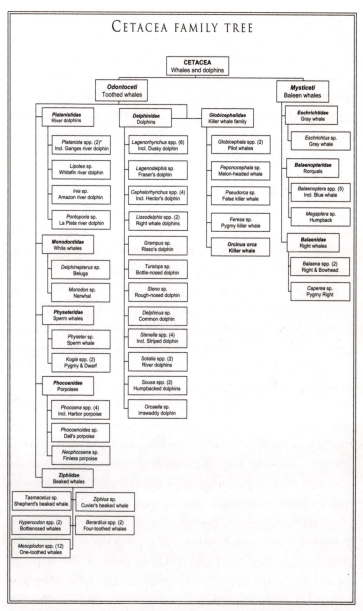

CETACEA FAMILY TREE

CETACEA
Whales and dolphins

Odontoceti
Toothed whales

Mysticeti
Baleen whales

Platanistidae
River dolphins

Platanista spp. (2)*
Incl. Ganges river dolphin

Lipotes sp.
Whitefin river dolphin

Inia sp.
Amazon river dolphin

Pontoporia sp.
La Plata river dolphin

Monodontidae
White whales

Delphinapterus sp.
Beluga

Monodon sp.
Narwhal

Physeteridae
Sperm whales

Physeter sp.
Sperm whale

Kogia spp. (2)
Pygmy & Dwarf

Phocoenidae
Porpoises

Phocoena spp. (4)
Incl. Harbor porpoise

Phocoenoides sp.
Dall's porpoise

Neophocoena sp.
Finless porpoise

Ziphiidae
Beaked whales

Tasmacetus sp.
Shepherd's beaked whale

Ziphius sp.
Cuvier's beaked whale

Hyperoodon spp. (2)
Bottlenosed whales

Berardius spp. (2)
Four-toothed whales

Mesoplodon spp. (12)
One-toothed whales

Delphinidae
Dolphins

Lagenorhynchus spp. (6)
Incl. Dusky dolphin

Lagenodelphis sp.
Fraser's dolphin

Cephalorhynchus spp. (4)
Incl. Hector's dolphin

Lissodelphis spp. (2)
Right whale dolphins

Grampus sp.
Risso's dolphin

Tursiops sp.
Bottle-nosed dolphin

Steno sp.
Rough-nosed dolphin

Delphinus sp.
Common dolphin

Stenella spp. (4)
Incl. Striped dolphin

Sotalia spp. (2)
River dolphins

Sousa spp. (2)
Humpbacked dolphins

Orcaella sp.
Irrawaddy dolphin

Globicephalidae
Killer whale family

Globicephala spp. (2)
Pilot whales

Peponocephala sp.
Melon-headed whale

Pseudorca sp.
False killer whale

Feresa sp.
Pygmy killer whale

Orcinus orca
Killer whale

Eschrichtidae
Gray whale

Eschrichtus sp.
Gray whale

Balaenopteridae
Rorquals

Balaenoptera spp. (5)
Incl. Blue whale

Megaptera sp.
Humpback

Balaenidae
Right whales

Balaena spp. (2)
Right & Bowhead

Caperea sp.
Pygmy Right

* (–) Numbers in parentheses denote number of species.

seems only distantly related to other Globicephalidae. Little is understood about the melon-headed whale, which was hardly known prior to the 1960s when a series of mass strandings brought it to scientific attention. The previous scarcity of this species is something of a mystery, for large schools of up to 500 are now known to be widely dispersed in all the warm and tropical waters of the world.

A more cohesive cluster of species is found among the pilot whales. These placid pod-living 'blackfish' are the dummies of the family. Whalers called them 'black sheep'—a reference to their single-minded adherence to their large pods. Like most of the *Globicephala* genus, the short-finned pilot whale is found mainly in tropical waters, while its close relative, the long-finned pilot whale primarily inhabits the cool temperate waters to the south and the north. Given its divided distribution, the long-finned pilot whale occurs in two distinct populations—North Atlantic and southern hemisphere—which are sometimes even accorded separate subspecies status.

While all other dolphins, indeed all other whales (both Mysticeti and Odontoceti), concentrated on the abundant non-mammalian prey the seas offered, the largest of the Globicephalidae preyed upon aquatic mammals themselves. Not only did this species consume the fish, squid and other prey that fed their relatives, but they also targeted their carnivorous rivals, the seals, and even their own cousins—the giant baleen whales. The 'whale-killers' had arrived.

Killer whales *Orcinus orca* soon spread throughout the oceans of the world, but they are most abundant in the rich, cold polar oceans, particularly of the southern hemisphere. These icy waters are rich in krill, which feed an abundance of baleen whales, as well as fish and squid, which, in turn, feed seals and penguins, all of which are eaten by killer whales. Huge super-pods of killer whales, numbering thousands, have been observed on the edge of the pack ice of Antarctica. Resident populations encircle the small outcrops of subantarctic islands where seals and penguins breed. Further north, the richness of the oceans gradually declines. The wide open oceans effectively isolate the Antarctic killer whales from their relatives further north, even from those as close as New Zealand and South America. Antarctic killer whales are larger than killer whales elsewhere and have a distinctive grey cape, rather than the crisp black-and-white characteristic of other populations.

Australia and New Zealand, on the northern fringes of the Southern Ocean, are also home to many seal colonies and hence killer whales. Killer whales are relatively common around New Zealand, but are less common in southern Australia. Further north, towards the warm tropical waters, killer whales are seen only occasionally.

Given this general distribution pattern, it is surprising to find that killer whales are more common along the east coast of Australia, with its warm, south-flowing currents, than along the cold, fish-rich southern coastline. There are few seals on the east coast, yet killer whales are rare but regular visitors, particularly from winter through to spring. These killer whales are not here for the fish, the seals or the penguins—they are following a bigger feast altogether—the baleen whales.

Baleen whales, like the right whale and humpback, migrate up the east coast of Australia every year to breed in the shallow warm waters of the Pacific islands. No doubt their smaller calves, which lack the great insulating mass of their parents, grow more rapidly in the warmer waters where they use less energy to keep warm. But the lack of food in these waters for the adults and the costs of the long migration must surely counterbalance this benefit. Smaller animals, such as seals and penguins,

THE WHALE'S CLOSEST RELATIVE

The evolutionary history of the whales has long been a topic of heated debate. Their mammalian characteristics gave them an obvious relationship with land mammals, but their highly evolved aquatic adaptations disguised many of the features that usually allow taxonomists to identify evolutionary origins.

Charles Darwin thought that whales might be most closely related to bears. The Carnivora family seemed to be the obvious candidate for the whale's closest living relatives—all whales are carnivorous and other carnivores, the seals, had adopted an aquatic mode of life. There were, however, other possibilities for whale relations. Some taxonomists noticed puzzling similarities between whales and pigs. Aquatic plant-eating Sirenia (dugongs and manatees) might be the closest living relatives of whales. The over-sized semi-aquatic hippopotomuses might also resemble primitive whales. But were these similarities due to a shared ancestry or to the convergence of body shapes as different species adapted to a similar aquatic lifestyle?

Modern molecular research does not suffer from the same problems of convergent evolution as external appearance. For example, sharks and whales look completely different at a molecular level, no matter how similar they might appear on the surface. Molecular taxonomy therefore has the potential to 'see' evolutionary difference and similarities between species that are not apparent on the surface.

continued . . .

. . . continued

Molecular taxonomy clearly places the whales as a sister group to the Artiodactyls (or ungulates like pigs, deer, camels and hippos). (Dugongs, for the record, are most closely related to elephants.) But within this group, whales seem to be most closely related to the hippopotamuses which may not be as closely related to other artiodactyls as once thought. Despite the current orthodoxy supporting a mesonchynid ancestor for modern whales, further research into the origins of both hippos and whales might reveal alternative ancestors, such as the pig-like anthracotherids.

which have even less thermal mass than baby whales, spend their whole lives in the cold waters. What benefits do the whales gain from this long and arduous voyage north away from their main source of food?

Some scientists have suggested that baleen whales migrate north to escape from killer whales. The hundreds of thousands of killer whales in subantarctic waters are a constant threat to baby baleen whales. Large baleen whales can store more blubber than smaller killer whales enabling them to go without food for longer periods of time. By moving north, the whales take their young to an area which—year-round—can support far fewer killer whales.

Only a few families of killer whales—perhaps resident in off-shore Australian waters, perhaps visitors from New Zealand or ocean-going transients—patrol the Australian coast along which the baleen whales, including humpback and right whales, migrate. And killer whales, although present, are even less abundant in their Pacific breeding grounds. Only a few highly

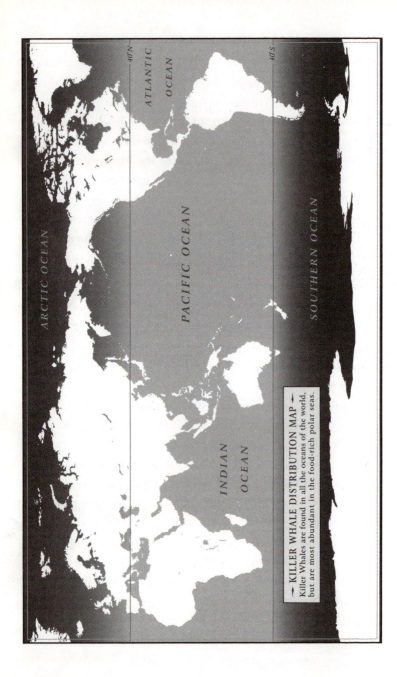

ARCTIC OCEAN

ATLANTIC OCEAN

40 N

PACIFIC OCEAN

INDIAN OCEAN

40 S

SOUTHERN OCEAN

← KILLER WHALE DISTRIBUTION MAP ←
Killer Whales are found in all the oceans of the world,
but are most abundant in the food-rich polar seas.

specialised and skilled whale-killers can afford the risks of following the humpbacks into this aquatic 'desert' where there are few seals or small whales to supplement their diets.

Killer whales are generally not regarded as migratory animals, although they do follow seasonal changes in prey abundance. The small proportion of killer whales specialising in large baleen whales are probably the most nomadic of all the killer whales. For example, while Antarctic killer whales generally only move north and south with the expanding and retreating icecap, following the greatest concentration of seals and penguins, conspicuous grey-caped Antarctic killer whales have been seen as far north as the Bay of Islands on the northern end of New Zealand. Killer whales hunting a Bryde's whale in California have been previously identified in Alaska, over 3000 kilometres to the north.

Before the devastation inflicted by human whaling, the scale of humpback and right whale migration along the east Australian coast must have rivalled that of the wildebeest in Africa, if not in number, then at least in bulk and spectacle. In the early 1840s Oswald Brierly described seeing large baleen whales (perhaps blue whales) in a huge school, filling the ocean as far as the eye could see. Such bounty would have supported far more killer whales than is currently the case.

Killer whales have probably been using Twofold Bay to trap migrating baleen whales for thousands if not millions of years, but the only killer whales we know about are the family that took up residence each year outside Twofold Bay from the nineteenth to the twentieth century. The Eden killer whales arrived in Twofold Bay every June, taking advantage of the nearby deep enclosed harbour to 'trap' the baleen whales as they travelled up, and later down, the east coast every winter and spring. This natural 'pocket' was large enough to drive baleen whales into, yet shallow and restricted enough to prevent them from escaping back to the ocean or diving out of reach. Many a baleen whale has been harried to death in Twofold Bay and many more driven ashore by the snapping pack of killer whales.

Despite the Twofold Bay's natural advantages the killers used it only for hunting. At other times, Eden residents observed the killer whales relaxing in Leatherjacket Bay, south of Twofold Bay and below one of the whaler's favourite lookouts, Boyd's Tower. Like the whalers above, the killer whales probably found this site offered them the earliest indication of baleen whales approaching from the south. Since killer whales use their hearing to detect prey long-range, Leatherjacket Bay may have provided an underwater auditory version of the view the whaler's sought from Boyd's Tower. But this exposed site may have had other advantages as well.

Killer whales often develop favourite 'lounging' areas. One beach on Vancouver Island, Canada, is a favourite haunt of local killer whales because of its smooth uniformly sized pebbles. These pebbles are an irresistible massage couch for the killers which rub themselves back and forth on the seabed. Similarly, the killers of the Crozet Islands spend their spare time loafing in a kelp bed of *Bay American*, apparently satisfying a similar urge for tactile stimulation. Film-maker Greg McKee speculated that Leatherjacket Bay might similarly harbour some kind of rubbing rocks or kelp beds which attracted the Eden killers during their spare time. When McKee dived in the bay in 1998, however, he found no geological or botanical features which might explain the killers' attraction to Leatherjacket Bay.

It is difficult to believe that Leatherjacket Bay offered more shelter from prevailing weather than the semi-enclosed Twofold Bay, but then perhaps 'shelter' is not what the killer whales required in their leisure hours. For a whale sleeping is a rather tricky procedure. Simply floating about might be safe enough in the open ocean, but close to shore might result in stranding. In these circumstances, killer whales perform a peculiar synchronised resting pattern. Clustering tightly together, the entire pod synchronises its breathing with the pace set by the lead female. Several short dives are followed by a longer dive of several minutes. Killer whales, like dolphins and some birds, seem to swim and sleep at the same time by resting first one half of their brain, then the other.

Locations for resting behaviour must, of course, be carefully selected. Not only must they be relatively safe from disturbances and dangerous conditions, but they must ideally be areas with strong and steady currents. Swimming against a moderate current enables the killer whales to stay in the one spot. And it is in Leatherjacket Bay that these steady currents can be found. The eastern current, which flows offshore down the eastern Australian coast, sets up counter currents closer to the coast. In the confines of Twofold Bay these currents are disturbed and eddy somewhat randomly around the bay, fuelled by winds and tide. But along the straight stretch of coast south of Red Point, including Leatherjacket Bay, the counter currents are more stable and predictable, commonly running at about one or two knots—the perfect speed for a leisurely cetacean sleep.

The waters around Eden at the turn of the nineteenth century might well have resembled a killer whale's ideal bed-and-breakfast. From their resting place in Leatherjacket Bay, the killers were perfectly positioned to spot any passing whale hugging the coastline. With the natural entrapment of Twofold Bay enabling them to drive their victim into the shallows, small wonder the killer whales returned to this area year after year to feed on the seasonal bounty. Eden was already naturally perfect; now all the killers needed was someone to help them kill their whales and Eden would indeed be paradise.

*C*lose nestled to her side was a youngling of not more, certainly, than five days old, which sent up its baby-spout every now and then about two feet in the air. One long, wing-like fin embraced its small body, holding it close to the massive breast of the tender mother, whose only care seemed to be to protect her young, utterly regardless of her own pain and danger. If sentiment were ever permitted to interfere with such operations as ours, it might well have done so now; for while the calf continually sought to escape from the enfolding fin, making all sorts of puny struggles in the attempt, the mother scarcely moved from her position, although streaming with blood from a score of wounds. Once, indeed, as a deep-searching thrust entered her very vitals, she raised her massy flukes high in the air with an apparently involuntary movement of agony; but even that dire time she remembered the possible danger to her young one, and laid the tremendous weapon as softly upon the water as if it were a feather fan.

So, in the most perfect quiet, with scarcely a writhe, nor any sign of flurry, she died, holding the calf to her side until her last vital gasp had fled, and left it to a swift despatch with a single lance-thrust. No slaughter of a lamb ever looked more like murder. Nor, when the vast bulk and strength of this animal was considered, could a mightier example have been given of the force and quality of maternal love.

Frank T. Bullen (1898), quoted in R. Mabey
The Oxford Book of Nature Writing, pp. 108–9

3

THE BALEEN AND BLUBBER BOOM

Twofold Bay ... we may venture to prophesy, will be one of the most important towns on the eastern coast of Australia. It possesses numerous advantages in itself, as well as forming a halfway house between Sydney and Van Diemen's Land; it is a safe and accessible harbour although confined; there is plenty of good wood and water in the immediate vicinity; there is also good land for pasturing ... it is situated in the most delightful latitude, so very picturesque in its adjacent scenery.

The *Australasian*, 3 June 1843, quoted in G. Waitt and K. Hartig, 'Grandiose plans but insignificant outcomes: The development of colonial ports at Twofold Bay, New South Wales', p. 201

In December 1842, a lean and elegant topsail schooner slipped into Twofold Bay. On the decks of the *Wanderer* stood a young man by the name of Oswald Walter Brierly. Brierly surveyed the picturesque bay with an artist's eye—the deep turquoise waters, framed by sandy golden beaches and ochre cliffs, the changeless and sombre olive tones of the gumtrees, the bleached white whale skeletons and the rudimentary bush-and-bark huts of the whalers and Aboriginal people. Trained as a naval architect and a graduate of Henry Sass's Art School in London, the young Brierly stood at the beginning of a long and distinguished career as a maritime artist. Twofold Bay was his first posting, albeit not as an artist, but as manager of a whaling station established by the Scottish entrepreneur Benjamin Boyd.

Strictly speaking, Twofold Bay lay beyond the 'Official Limits of Location' designated by Governor Darling in a vain attempt to control settlement in Australia. But a port like Twofold Bay, with so many natural resources and smack in the middle of the trade route between Sydney and Van Diemen's Land, was irresistible to the squatters and settlers, entrepreneurs and escapees who were establishing settlements wherever the economic circumstances seemed favourable. The embryonic government was forced to stretch its resources to administer such outposts or risk having them operate outside its jurisdiction entirely. The port of Eden was born.

In 1842, Eden was the centre of an empire founded by the three Imlay brothers, Peter, George and Alexander a decade earlier. The Imlays had taken up vast stretches of land between Bega and the present Victorian–New South Wales border, channelling their produce down to the port of Eden, where they also ran a whaling station. Whaling had been successfully conducted from Twofold Bay since 1828, when Thomas Raine whaled for a season. But whaling is a notoriously fickle and unsustainable industry and most whaling activities last only a season or two. The Imlays, despite their vast empire, proved no exception and, by 1842, many of their whaling assets were being taken over.

Nevertheless, Brierly's employer, Benjamin Boyd, saw an opening for himself in Twofold Bay. Backed by wealthy London financiers, and well-connected enough to warrant a twelve-gun salute on his arrival in Sydney, Boyd wanted to establish his own pastoral and whaling empire. Not content with buying up large tracts of cheap land at Eden (with the excessively generous assistance of local officials), Boyd decided to establish a rival town of his own on the opposite side of the bay—Boydtown. Brierly was to be installed as manager of the whaling station at East Boyd. He stayed over Christmas and into January of the following year, obtaining information from the local whalers in preparation for the winter whaling season ahead.

Like many natural history and landscape artists of his era, Brierly's artistic talents were combined with a sharp eye for detail and a mind for shrewd scientific observations. He engaged the eminent British scientist Sir Richard Owen in debates, summarily dismissing Owen's claim that the 'great whalebone whale' (right whale, *Balaena glacialis*) was restricted to the northern hemisphere. He wrote extensively on the whales of the southern hemisphere, which had only recently come to scientific attention. It is from Brierly's journals that we have our first written account of the activities of killer whales in Twofold Bay.

Even before he had embarked on any whaling activities, Brierly had heard about these remarkable animals. His information probably came from Aboriginal whalers who may have been the only whalers resident in Twofold Bay over the non-whaling summer season. Brierly noted the close relationship of the Aboriginal people to the killer whales:

> One of their number was knocked out of the boat by a blow from the tail and killed. This frightened them very much for they imagine that when any of them die his soul goes into the body of a black fish and previously that day they said/saw that the fish was a gin belonging to the Gomah Beach blacks with whom they make war (*Journal of a visit to Twofold Bay Dec 1842–Jan 1843*, MLA535: frame 0276).

By the middle of the next whaling season, Brierly knew about the killer whales' role in the bay's whaling activities and regularly wrote of their involvement. Every whale capture in his diary is marked by a characteristic whale-tail sketch in the margin.

> Sunday 7th of August, 1843…the killers are on him, surrounding and darting at his tongue when the pain of his wounds cause him to open his mouth. This fish is a species of porpoise from eighteen to twenty feet long. They attack the whale in packs and seem to enter keenly into the sport; rising and plunging about the boat—generally preventing the whale from escaping his captors and meeting him at every turn— tearing away portions of the lip and tongue while still alive…[The whale dies with] one tremendous shudder…In a few minutes the killers took him bodily down. Large pieces of the lip torn off rising to the surface showed that a regular feast was going on below. The whale generally floats in 24 hours when the killers have devoured the tongue (*Diaries at Twofold Bay and Sydney 1842–4*, MLA539: frame 324).

Brierly was impressed by the activities of the killer whales, frequently commenting on their skill and value to the whalers. He noted, with some incredulity, that the Aboriginal whalers 'pretend' to be able to identify them individually as reincarnated spirits of their former comrades. His sketches of whale hunts (as shown on the chapter opening pages of this book) and etchings regularly feature a handful of killer whales cavorting around the dying whale. Although Brierly does not mention the number of killer whales involved in these hunts, unsubstantiated reports have claimed there were 'hundreds' at this time.

Boyd's empire, like the Imlay's before him, was short-lived and by the time Brierly left Twofold Bay after just three seasons, Boyd's days were numbered. In 1847, Captain Owen Stanley visited Twofold Bay to survey the South Head Light House and befriended the increasingly frustrated Brierly. Stanley persuaded Brierly to join him on a survey of the Great Barrier Reef, Louisiade

EVENTS AT EDEN TIMELINE

Year	Event
1797	Twofold Bay discovered by George Bass on 19 December
1798	Twofold Bay named by Bass when sheltering in *Tom Thumb* at Snug Cove Bay surveyed by Bass and Flinders in the sloop *Norfolk* in October
1828	Thomas Raine of Sydney begins whaling
1834	Peter, George and Alexander Imlay begin whaling
1842	Ben Boyd arrives in Australia
1842–3	Oswald Brierly visits Twofold Bay between December and January
1843	Boyd's first whaling season from hulk at East Boyd Bay Alexander Davidson moves to Boydtown
1843–4	Oswald Brierly appointed manager of whaling in Twofold Bay
1845–8	27 open boats operating with both Aboriginal and European crews out of East Boyd, Kiah Inlet and Snug Cove
1846–8	Brierly leaves Twofold Bay
1847	Imlay whaling taken over by Walker Boyd ceases whaling operations
1860s	Alexander Davidson begins whaling with son John
1864	George Davidson born
1878	George Davidson begins whaling
1890s–1920s	Davidson family the only whalers in Twofold Bay
1929	No whales processed in Twofold Bay after 1928

Archipelago and New Guinea and on 18 April 1848, Brierly left Twofold Bay. He returned to New South Wales in December of the same year before leaving for England via the Pacific on the *Meander*. Commissions as a war artist in the Crimea and for the royal family were soon followed by world cruises with the Duke of Edinburgh, one of which, on the *Galatea*, brought Brierly back to Australia's shores twenty years after his last sojourn. An appointment as marine painter to Queen Victoria and to the Royal Yacht Squadron and then a knighthood saw Brierly well set up in his mature years and far from his early days as a humble manager of a rough-and-ready whaling station in the raw Australian colony. But throughout his career, Brierly continued to draw upon his early experiences as a whaler and frequently reminisced about the strange relationship between the whalers and killer whales of Twofold Bay:

> In these waters there is often found in company with the whales a species of large porpoise with blunt head and large teeth. This animal is known as a killer. These killers will frequently attack a whale and at times succeed in worrying it like a pack of doges. The whale men are very favourably disposed towards the killers and regard it as a good sign when they see a whale hove to by these animals—because in that case they regard it as more likely to find an easy prey in the whale when assisted by their allies the killers. When a whale has been killed by the whaling party it is no uncommon occurrence for the killers to take the whale down bodily—when this happens the only thing to do is to attach a buoy to mark the place after being down some 24 hours the then inflated carcass usually rises—it is then found that the killers have eaten the whale's tongue—no doubt this delicacy is regarded by them as their perquisité (*Reminiscences of the Sea*, MLA546: frames 9–10).

Just as Brierly left Twofold Bay, Boyd took on a new employee, Alexander Davidson, whose family was to prove an exception in the transient and unstable world of shore-based whaling. The Davidson dynasty would whale the waters off

BAY WHALING

Whaling began as a shore-based activity in many coastal communities. Hunting whales from small boats along the coast is a traditional part of many cultures, from the Atlantic to the Pacific. Typically small boats are used to approach the whale, which is then harpooned.

Commercial bay whaling, or shore whaling, expanded its range with the colonial expansion of European nations around the world from the 1500s onwards. Whalers and sealers were often the first European visitors to foreign shores—setting up short-term processing plants onshore before returning to Europe or America loaded with oil. When whales were sighted offshore from prominent lookouts, open boats manned by half a dozen men rowed off in pursuit. Approaching the whale, a man in the bow would harpoon the whale before racing aft to take up steering. The boat steerer, in charge of the vessel, then moved forward to lance the whale, delivering the death blow. The whale could then be towed to shore for processing.

Twofold Bay for over 70 years. The longevity of their whaling operation in the face of diminishing returns was probably due in no small part to the assistance of the killer whales.

Alexander Davidson arrived in Twofold Bay as a craftsman for Boyd, but by the 1860s he had begun whaling. His son, named John, followed in his father's new profession establishing a whaling station at Kiah Inlet. The family built their home near the trying-out works out of the wreck of the barque *Laurence Frost*. By 1864, a third generation of Davidson whalers was

OFFSHORE WHALING

According to whaling legend, offshore whaling began in 1712 when by a Nantucket coastal whaler called Christopher Hussey was blown out to sea in his small boat and managed to harpoon a sperm whale. Whaling in open waters from larger boats soon became profitable. No longer were whales safe from humans in the vast open oceans that had long been their haven, but were pursued to the ends of the earth, fulfiling an insatiable demand for oil and other whale products. Offshore whaling was particularly useful in hunting sperm whales, which rarely come close inshore and which yield high-quality oil levels. By 1793, several whaling ships operated off the east Australian coast. Whaling ships were often away for up to three years and their crews tended to be a motley collection of outcasts and misfits of all nationalities.

Ship whaling, or offshore whaling, followed similar traditions to shore-based whaling, with whales being pursued by four or five open row boats launched from the mother ship. After capture, the whales were processed on board the larger craft. Being able to follow the seasonal movements of the whales significantly increased the capture rates of offshore whaling compared to bay whaling. Factory ships, which could process large numbers of whales, ensured that the industry rapidly outstripped sustainable harvest levels, resulting in whale populations which are now only a fraction of their pre-whaling size. While bay whaling was limited in its impact on whale numbers, offshore whaling has had a far more significant impact.

born—George Davidson—who started whaling in his father's boats at the tender age of fourteen. George later recalled that when he began whaling, there were about 27 killer whales operating in the bay, most of which could be individually recognised by their uniquely shaped dorsal fins. Hooky, Stranger and Cooper each led a sub-group within the pod. George also recalled that a mischievous young male killer by the name of Tom was also present at the time.

By the end of the 1800s, the killer whales' activities had attracted the attention of the press. An 1872 article in the *Illustrated Sydney News and New South Wales Agriculturalist and Grazier* outlined in detail the assistance the killers provided to the whalers, noting that the killers 'take it [the whale] to the bottom (if not driven away by the whalers) and feast upon the throat, lips and tongue of the dead monster'. An 1895 article in the local *Bega Standard* briefly described the killers' cooperative activities. A newspaper account of 14 June 1898 read as follows:

Eden, Tuesday.—Early on Sunday morning two large whales were observed in conflict with the 'killers' in North Bay. Two whaleboats came up with the whales when off the light-house, and immediately made fast to them. The whales went off at great speed up the harbor, then, turning round, went out round North Head to Kokariracki, where one was killed, while the other, doubling back, was killed off South Head. The steersman of one of the whale-boats was twice jerked overboard during the sudden rushes of the whale, but quickly climbed back again. The chase was witnessed by a large number of people. One whale was towed to the whaling station the same evening (as quoted in R. Davidson, *Whalemen of Twofold Bay*, p. 104).

But it was not until the killer whales had begun to abandon Eden that a need to record this remarkable episode in Australia's whaling history became more urgent. Between 1900 and

Davidson's whaling station.

1950 there was a rush of information recorded on the killer whales—first-, second- and third-hand accounts, personal observations, recollections and reminiscences, tall tales and local mythology. As the number of killer whales returning to Twofold Bay each year declined, the information and stories about them increased, becoming more and more shrouded in myth and exaggeration.

Both contemporary and retrospective articles about the killer whales suggest that about twenty to 30 killer whales returned to Eden every winter from the 1840s until 1901. A 1903 article in the *Sydney Mail* reported that 'When the whale is passing north it is driven into Twofold Bay by what are known as Killers ... Any attempt the whale makes to go out to sea the Killers resent with all energy by snapping pieces out of it', but makes no mention of the number of killers involved.

For an industry steeped in the deaths of countless whales, it may have been a single death—that of Typee—that triggered the beginning of the end for the killer whales' association with Eden. According to Charles Welling, in the winter of 1901, the pod was pursuing a small 'grampus' whale near Aslings Beach to the north of Eden. It was their habit to drive their prey into shallower water, cutting off any hope of escape. In his enthusiasm, and with a retreating tide, Typee stranded himself in the surf on the beach and was unable to make his way back to deeper water. Such semi-intentional stranding was not entirely uncommon. Often the killer whales managed to flip themselves back into deeper water. Occasionally they failed. One of the older killer whales, Big Ben, died after becoming stranding on rocks near Leatherjacket Point.

George Davidson happened to be taking a boat out to inspect a buoyed whale and saw Typee's predicament. His crew rowed frantically to rescue the stricken killer whale, but not before a man on the shore had made his way down to the whale. To the horror and distress of the whalers and killer whales witnessing events from deeper water, he drew out a knife and plunged it into the killer whale. Apparently, he intended to sell the skeleton to a museum. Instead he drew the wrath of the local whalers, whose reaction, armed as they were with harpoons and tomahawks, convinced the culprit to quit the scene before they reached shore. The local constabulary soon tracked him down and advised him to leave town quickly and quietly that night for they could not guarantee his safety against any of the whalers, particularly the outraged Aboriginal crewmen. The killing of dolphins is taboo for most coastal Aboriginal communities, and in Eden, the killer whales were sacrosanct.

The Aboriginal whalers, with their strong spiritual association with the killer whales, must have viewed the deliberate slaying of this animal as a serious misfortune. Grave repercussions would certainly have been expected from the spirit world for this death. Not long after Typee's death, the Aboriginal whalers' permanent camp at Kiah Inlet was abandoned and the tribe moved north to Wallaga Lakes, leaving George Davidson with a chronic shortage of labour for the whaling station.

The killer whales seemed to have felt the death of their fellow keenly as well, for reports state that only a handful returned the following year, with most of the pod never being seen again. Killer whales live in tight-knit, long-lived family units and a death often seems to disrupt the usual behaviour of family members. John Ford, who has studied the killer whales of British Columbia for over fifteen years, recalled that when an elderly female died in one pod, her two adult sons left the pod to travel alone before returning to their family six months later. Perhaps a similar family bond explains why Young Ben is said to have left Twofold Bay after the death of Big Ben.

There are many reasons why the killer whales may have begun to abandon Eden as a winter hunting ground—the declining number of baleen whales, lower numbers of seals and other small whales to sustain them along the coast in the absence of baleen whales, or even the activities of offshore whalers. But the deliberate slaughter of one of their family may well have made Twofold Bay seem dangerous as well as unrewarding.

By 1902 only seven killer whales were returning to Twofold Bay each year, including Tom, Hooky, Humpy, Young Ben, Kinscher and Charlie Adgery. By 1923, only three of these animals returned. Tom and Hooky continued returning until 1928, the last year an obviously aged Hooky was seen.

The killers were not the only whales in decline. Their prey had also been drastically reduced in number, particularly by modern ship-based whaling methods. Some species, like the right whale, had been reduced to near extinction. The remaining whales were both fewer in number and smaller in size. In

A whale boat of Twofold Bay, with baleen whale calf and killer whale.

Brierly's time as many as eight whales could be caught in a day. Typically a whale a day was caught for most of the season. By the end of the nineteenth century, for the Davidson's station twenty whales for the whole winter was a good season.

It was the advent of alternatives to whale oil that tolled the final death knell for the Twofold Bay whaling industry, however. Coal gas had already been reducing demand by the 1820s, with petroleum and kerosene further removing market share from the 1850s onwards. Modern oils, such as jojoba, have completely replaced whale oil in many products, including machinery, phar- maceuticals and cosmetics.

As the whaling industry drew to a close, so too did the regular presence of the killer whales. Tom was the last of the Eden killer whales to be seen. After Hooky's presumed death, he returned alone each year to the bay. He was an old killer whale by then with stubs of teeth worn down through a lifetime of predation. The infectious energy with which he had approached the hunt for most of his life was much diminished. Baleen whales would probably have been beyond him, even if there had been baleen whales to catch. But no baleen whales had been caught off the bay since 1926. Tom probably even found the dolphins, seals and small whales which now comprised his main diet difficult to catch on his own. Just days before his death he was apparently seen harassing a minke whale, but when George Davidson retrieved his dead body from the bay in September 1930, Tom's stomach was empty. The baleen and blubber boom was well and truly over for whalers and killer whales alike.

*L*one handed, George Davidson, a 70 year old master whaler, attacked and killed a humpback whale at Twofold Bay. The veteran whaler worked from a small dinghy and used only a lance. Although whaling at Twofold Bay ceased with the disappearance some years ago of the whaler's allies, the famous pack of killer, interest in that adventurous pursuit has been revived by Davidson's achievement.

For some weeks whales southward bound had been cruising in and out of the bay and although men and boats were not available, the veteran man went to sea alone in his dinghy from the old Kiah whaling station armed with a lance to attempt to destroy a whale that was in the harbour. The possibility of using a boat gun and bombs was over-ruled because of the risk that the dinghy would overturn with the force of the discharge of the gun.

Having sighted the whale from a headland, Davidson rowed cautiously to the attack. As he neared the whale it sounded. He moved to where he expected it to re-appear, and as it rose he drove the lance deep into a vital part of the body. The whale, with a mighty plunge, sounded deeply. On coming to the surface it appeared to be distressed but swam out of the bay. Davidson followed, convinced that the whale had been mortally wounded, but it out-distanced him. Four days later its body washed up on rocks near the light-house. Attempts to float it off have so far been unsuccessful. Davidson's feat of attacking and killing a whale without assistance is unparalleled in the history of local whaling.

Undated newspaper clipping 'Lone whaler: Veteran in dinghy',
circa 1939 (extract from *Whalemen of Twofold Bay*, p. 128)

4

OLD TOM

The killer known as Tom was extremely playful, if not actually mischievous, and constantly gave trouble to the whalers because of his habit of seizing the ropes and running off with them. The same, it is alleged, applied to mooring lines from fishing boats. Tom would pick rope and kellick {anchor} up and make off with them, much to the discomfiture of the fishermen, who were at times placed in great danger of drowning from this unpleasant habit.

D. G. Stead, *Giants and Pigmies of the Deep*, p. 42

Tom developed a particularly close relationship with the whalers of Eden. But his loyalty to Eden is no more unusual than that of numerous other local cetacean celebrities around the world which, for some unknown reason, become particularly attached to specific locations and even seem to seek out the company of humans. 'Pelorous Jack' was a solitary Risso's dolphin who lived in the Marlborough Sounds of New Zealand for over twenty years, seeking out the company of local sailors and fishermen. In many ports around the world there are stories of lone dolphins, perhaps injured or sickly, perhaps just outcasts from their own kind, which take a particular interest in humans. When I lived in Port

Adelaide, a local horse trainer swam the horses behind a dinghy for exercise in the Port River. For many years, the same solitary dolphin accompanied their early morning constitutionals.

The reason why such generally social animals are found alone is a mystery, but their solitary status may well explain their strange interest in humans. Tom was probably the last member of his immediate family and may have returned to the familiar surroundings of Eden for the company of his erstwhile hunting partners. Tom, like Pelorous Jack and many other cases of solitary dolphins, could be explained away as an idiosyncratic individual—an outcast—who, deprived of the company of his own kind, took solace in the company of humans. But unlike all those other local cetacean celebrities, Tom was not originally a social outcast. He was simply the last of a large family of animals, all of whom exhibited a similar interest in humans. And the behaviour of a whole clan of killer whales cannot readily be dismissed as a one-off aberration.

Tom's status as the last of the killer whales is not the only factor that has immortalised him in local memory. Even when alive, Tom's behaviour marked him out as a great 'character'. Along with Humpy, Charlie Adgery and Old Ben, Tom was one of the more athletic and skilful whale hunters. The recollections of Margaret Brooks (recorded by Greg McKee), who accompanied her father John Logan on his launch to observe some of the last whale hunts in Eden, suggest Tom's enthusiasm remained undimmed even in his declining years. Gripping tenaciously onto fluke and fin, Tom would be hurled bodily out of the water, sometimes feet into the air, as the hunted whale sought to dislodge its tormentors. Even crashing bodily back into the water failed to force Tom to loosen his grasp. Nor was Tom's gusto entirely restricted to his 'work'. His zest for life was also apparent in his leisure hours.

Tom was first recorded in the literature by E. J. Brady in 1909. Brady described the Eden pod as including 'Old Humpy, Stranger (the leader of the mob), Jackson, Kinchen, Tom—the humorist—Cooper, and Hookey' (p. 40). Subsequent accounts provided more detailed descriptions of Tom's 'sense of humour'. Charles Wellings related how 'Tom developed a bad habit of getting across the slack line between the whale and boat, holding it under a fin and towing the boat in any direction. This habit was deliberate, and appeared to be mischief, though a nuisance' (p. 292).

The whalemen were not the only ones 'blessed' with Tom's attentions. He seemed to enjoy towing as much as being towed. Wellings noted that 'even small boats whilst fishing have had the kellick [anchor] line taken and the boat dangerously towed off' (p. 292). Leatherjacket Bay, the killer's seasonal encampment, was particularly unsafe for fishermen. Hawkins & Cook reported that several fishermen 'found themselves being towed to sea at a rate not at all acceptable' (p. 271)—often swamping the small boats. Margaret Brooks recalled that her father, John Logan and Tom literally came to blows over this habit in later years. Tom revealed his displeasure at Logan's efforts to dislodge him with a boathook by attempting to fling himself onto the back of the boat, just as he did with the baleen whales during the hunt.

Tom's fascination with mooring lines was seized upon by local storytellers and led to the claim, promoted even within the Eden Killer Whale Museum, that Tom wore down his teeth by grasping the harpoon line. This tale seems to be supported by the fact that half a dozen of Tom's front teeth, top and bottom, are worn to stubs of just a centimetre or two. His back teeth are in slightly better condition, but the penultimate tooth before the front stubs has a curious semi-circular groove gouged out. This strange

Tom's skull (in the Eden Killer Whale Museum) shows the tooth wear that gave rise to the tale.

pattern of tooth wear lends credence to the idea that a rope wore Tom's teeth away.

The truth, however, is more prosaic. It is not even clear if Tom did grasp any lines in his mouth. Some reliable accounts like Charles Wellings's suggest he 'hooked' them over his fin instead (a behaviour often observed in captive dolphins and killer whales). George Davidson's grandson, René, has long been suspicious of the tooth-wear stories which formed no part of his family's oral history. After many years of searching, René finally obtained photos of rare killer whale skulls from the Monaco Museum which reveal exactly the same tooth-wear pattern as Tom's, even to the point of being more worn on the left than the right side of the jaw. Like most mammalian predators, killer-whale teeth are not replaced during their lifetime, and simply wear down to stubs. The centimetre or so of remnant stub visible in Tom's skull would have been flush with his gums in life. Similarly the semi-circular groove, so suggestive of rope wear, halfway along his rear tooth, is actually a cavity developed at the gum line and would have accelerated the loss of teeth.

Despite the dearth of scientific information on the killer whale before the 1960s, dozens of scientific papers and books have reported this peculiar pattern of tooth wear in the species. By the late 1800s, killer-whale tooth wear was thoroughly documented by luminaries such as J. E. Gray and Richard Owen of the British Museum and Danish professor Daniel F. Eschricht of physiology and anatomy at the University of Copenhagen. Even dentists studying their profession under the tutelage of Sir Charles Tomes's classic, *A Manual of Dental Anatomy*, were informed that the 'teeth of the grampus (Orca) become worn down on their opposed surfaces', often leading to exposed pulp cavities and abscesses. Yet despite the abundance of scientific evidence confirming that Tom's teeth were nothing unusual, local mythology maintains the rope-wear story.

The infection and abscesses alluded to by Sir Charles Tomes may occur if collected food is not removed by further biting and shearing action. Tom's skull shows evidence of just such an abscess on the upper right side. Some commentators have speculated that this abscess and subsequent blood poisoning or brain infection may have contributed to Tom's death. Whether the immediate cause of Tom's death was blood poisoning or starvation through worn teeth, 'old age' was ultimately responsible. Tom had been known and recognised in Eden for the past 50 years, and some claimed he was closer to 90 years in age when he finally died.

When the eminent Australian marine biologist, William Dakin, investigated the story of the Eden killer whales in 1932, he accepted local accounts that Tom was 80 to 90 years old when he died. Dakin was initially sceptical of the entire Eden killer-whale story, but after interviewing numerous individuals involved in whaling, many of whom had long since left Eden, he was struck by the consistency of their stories and descriptions. In many cases, Dakin noted that the events were recounted in a simple matter-of-fact manner without any hint that the participants considered their experiences strange or unusual. The last trace of Dakin's scientific scepticism was overcome by his discovery of Oswald Brierly's diaries, with their descriptions of the killer whales performing the same behaviours so many decades earlier. As Dakin pointed out, Brierly took his diaries to England with him. The diaries only returned to Australia when they were purchased by collector David Scott Mitchell, presumably after Brierly's death in 1894. The Brierly diaries subsequently formed part of the substantial Mitchell collection, which was bequeathed to the State Library of New South Wales in 1907. The diaries were probably not available to the public when articles about the Eden killer whales were written in 1908–9, nor were they likely to have been referred to by the whalers and local identities whom Dakin interviewed about the killer whales in the early 1930s. The independence of accounts by whalers in the early 1900s from Brierly's accounts of the early 1800s is further confirmed by significant differences in detail. Brierly never names individual killer whales, whereas individual names feature prominently in later accounts. And while Brierly emphasises the Aboriginal people's relationship with the killer whales, this connection is rarely made in later years. Both, however, confirm the basic features of the cooperative hunting arrangement—and for Dakin corroboration of the whaler's stories with a previously unknown historical account was confirmation of their veracity.

Having concluded that so much of this fantastic tale was probably true, Dakin saw no reason to discount local estimates of Tom's age. The only problem with the local estimates of Tom's age is that they exceed the maximum age to which male killer whales are thought to live by up to 40 years.

The whalers of Eden regarded all of the killer whales as males—perhaps reflecting the gender bias in their own ranks. In the last 30 years, however, detailed studies of killer whale pods

MALE AND FEMALE WHALE SIZES COMPARED TO TOM

Characteristic	Male	Female	Tom
Average length	6.7–7 m	6–6.5 m	6.7 m
Average weight	4–4.5 tonnes	2.5–3 tonnes	Unknown
Average height of dorsal fin	2 m	1 m	1.72 m

in British Columbia and elsewhere, reveal that they are typically matrilineal—consisting of a mother and her immediate family. The older females tend to lead the pod, although males may lead sometimes. Hunting is very much the remit of both sexes and, although many of the photos of Eden whales are of males, there must also have been females in the pod. As noted earlier, small finned whales like Kinscher and Sharkey were almost certainly females. The diminutive name for Little Jack may also have been indicative of a smaller female, while seniority and longevity within the pod is suggestive of a female gender for killer whales like Stranger. So, despite his name, could Tom have been female, thus accounting for his apparent age?

Some writers have suggested that Tom was unusually small for a male killer whale. The literature on killer whales commonly cites lengths of up to 9 metres for adult males—although this may be based as much upon hearsay and old whalers' tales as actual scientific evidence. The skeleton in the Eden Whale Museum is of an animal just less than 7 metres long, well short of the reputed maximum size. Killer whale dimensions can, however, only be properly assessed from captured whales, and killer whales are not particularly common captures. Live orca captures in North America tend to be of smaller, more readily transported animals. Killer whales have, however, been culled in Norwegian waters where they threatened the local fishermen's monopoly on herring stocks. Measurements of these whales provide a more reliable estimate of body size.

The largest killer whale measured in the Norwegian study (out of about 200 animals) was just over 7 metres long. The most common adult male body size was just over 6 metres—almost exactly the same size as Tom. Although killer whales caught in Antarctic waters tend to be up to 1 metre longer than Norwegian ones, local killer whale researchers like Ingrid Visner state that Australian and New Zealand killer whales are similar in size to the Norwegian animals. The widely cited figure of '9 metres' is probably over-stated, and a 6-metre killer whale is a perfectly reasonable length for a male.

Photographic and historical evidence also confirms Tom's masculinity. Tom was one of the best photographed of the Eden killers, with shots revealing a long, pointed dorsal fin characteristic of an adult male. Written descriptions of Tom also refer to his particularly high fin with an indentation near the top. These observations were confirmed too when George Davidson measured Tom after he died, revealing the enlarged dorsal and pectoral fins typical of male killer whales.

With Tom's masculinity fairly well established, can his age be verified? There seems little doubt that Tom was an old whale when he died, at least judging from his worn-down teeth. However, Tom's teeth reveal something else. In 1977, zoologists Edward Mitchell and Alan Baker persuaded the curators of Tom's skeleton to part with a single tooth for analysis. A cross-section of Tom's tooth revealed about 25 layers known as 'dentine growth-layer groups', which many researchers suspect roughly correlate with years. Norwegian studies, for example, have found between nineteen and 28 growth-layer groups in male killer whales of Tom's size. Baker and Mitchell also recorded a further ten outer cementum layers on Tom's tooth causing them to make a final estimate of Tom's age at about 35, rather than the 80 to 90 years of popular wisdom.

If Tom died at the age of 35 in 1930, he would have been born only in 1895, just a few years before the killers began to abandon Eden. The earliest date at which he could have reached maturity (and attained the distinctive dorsal fin by which he was recognised) would have been about 1902. And yet by 1909, E. J. Brady listed Tom as one of the best known killer whales. Furthermore, it is arguable as to whether Mitchell and Baker should have included cementum layers as 'years'. Most scientists assume that the dentine growth-layer groups alone should be read as years. This would put Tom's age at 25, suggesting he was merely an infant of four years and a long way from maturity when Brady named him in his article.

By the early 1900s, George Davidson, who had the closest relationship with Tom, was in his forties and had been whaling for over twenty years. Yet George Davidson asserted that Tom had been present when he first began whaling in 1878. While some room must be made for errors of memory, it is unlikely that Davidson would have erred to the tune of over two decades. If Davidson's memory serves us correctly, Tom would have died when at least 52 years old, and probably closer to 60 if we allow a few years for him to reach maturity before being identifiable and actively hunting with the pod. This estimate is certainly old for a killer whale, but is within the maximum age generally assumed for wild male killer whales.

There are many possible reasons why the layers in Tom's teeth reveal a younger animal than local history suggests. One reason may be that the skeleton that stands in the Eden Killer Whale

Museum is not the Tom of local legend at all. 'Tom' may have been more than one individual, with an old whale from early times replaced by a younger whale in later years. Dakin thought there was little chance of such false identity in the latter years of Tom's life. He argued that 'for over fifty years he [Tom] has been watched at such close quarters as would make mistake impossible'. A more likely source of error may lie in the scientific estimate of dating by dental growth layers.

Measuring the age of wild animals is notoriously difficult. Scientists estimate the age of wild killer whales in British Columbia from the number of sons and daughters associated with a mother multiplied by the average length of time usual between births plus the time a female killer whale takes to reach sexual maturity. The only truly reliable method is to identify a particular animal at birth and follow its progress through life until it dies. Few scientists have the resources to study the same individuals for so long. When individual animals live as long as or even longer than humans, only research over several generations of scientists will be able to deliver a clear answer on longevity.

A shortcut to estimating age is therefore one of the holy grails of zoological research. And teeth offer the promise of just such a shortcut, albeit only after an animal has died. Unfortunately, the promise rarely lives up to the reality in many cases. Clearcut dentine (or bone) layers appear to correlate reliably with years only in animals with strong seasonal patterns of growth. Hibernating animals, for example, are good candidates, where bone deposition effectively ceases for several months of the year. Animals with a constant supply of food year-round are less likely to exhibit such clear annual layers, indeed some layers may coincide with irregular periods of food abundance, while a lean year or two may result in no growth layers at all. Without a calibration of dentine layers from killer whales of a known age, we have no way of knowing whether or not they actually predict killer whale age in wild animals.

Other clues about Tom's age come from the origin of his name. Tom, like so many other killer whales, was probably named after a deceased Aboriginal whaler, following the local tradition that dead warriors 'jumped up' as killer whales to assist their relatives in the whale hunt. The death of a whaler called Tom might provide some substantiation of Tom's lifespan. Tom may have been named after Tommy the Stockman (Marimbine) whose accidental death in his early fifties is attributed to a boxing match with a kangaroo. We cannot be sure, however, that Marimbine was ever a whaler. Sue Wesson's research suggests that a far more likely namesake is the famous whaler, Tommy Headsman (Womanlin). Womanlin was born in either 1821 or 1822 at Snug Cove and probably died in the 1870s, around the time when George Davidson identified Tom in the pod. If this nominal lineage is correct, Tom was more than just a helpful animal for the Aboriginal community in Twofold Bay—he was a member of their family, the reincarnated embodiment of the soul of Womanlin.

Tom's apparently close relationship with the Aboriginal whalers extended into the white community, particularly the Davidson family. Tom was one of the whales that entered Kiah Inlet to 'call' out the whalers. According to interviews with George Davidson it was Tom who 'stood by' Davidson when he was knocked overboard and had to wait for his boat to return and pick him up. Tom was also one of the killers Davidson and the other whalers used to feed fish to, while waiting for whales. In his later years, feeding Tom may have become even more of a favoured pastime as the number of whales, and killer whales, declined. The Davidsons were particularly interested in Tom and Tom seemed to reciprocate this interest in the Davidsons.

Today, a road cuts through the dense forests surrounding Twofold Bay into the old whaling station at Kiah Inlet, but when the whaling station was active the only access was by sea. And despite the generally protected nature of Twofold Bay, the crossing from Eden to Kiah Inlet could be hazardous. On 24 October 1926, George Davidson's son Jack was returning from a cricket match in Eden with his wife Anne and three young children, Roy, Marion and Patricia. Despite having been a fine day, a storm out to sea had pushed a heavy swell into the bay and the shoaling mouth of Kiah Inlet swept the waves higher. As they approached the inlet one such wave caught the boat on its side, capsizing it. Jack and two of the children were swept away, leaving Anne and Marion clinging to the upturned boat until rescued by the Davidsons.

Interviews with family present at the tragedy, particularly John's sister Elsie, who rowed the rescue boat, reveal that while searching for the bodies, the Davidsons and other residents were joined by Tom who swam around and around the area where the dinghy had capsized. The children's bodies were soon recovered but it was several days before Jack's body was finally located, partially obscured on the sea floor in the area where Tom had been persistently swimming. Tom accompanied the whaleboat with Jack's body back to the wharf at Eden before heading out to sea, leaving George to wonder at Tom's apparent interest in their family tragedy and whether the killer whale had been trying to tell them where the body was.

No doubt Tom had been aware of Jack's body on the sea floor. How he interpreted that information is open to speculation. Humans in the water often seem to fascinate killer whales. Tom's behaviour in circling Jack's body may simply have been the result of curiosity. Accompanying the boat carrying the body back to the wharf, however, suggests more than idle curiosity.

Cetaceans instinctively assist their own infant, injured or

ailing family members to the surface to breathe—a behaviour commonly observed and exploited by whalers. Sometimes this behaviour extends to other species. Scientists have seen dolphins supporting dead seals, fish and birds. A young killer whale was even observed supporting an injured seal pup that would normally constitute its dinner. Such apparently misguided rescues probably reflect the strength of the instinct to protect family members from drowning. The notion that dolphins help drowning sailors by lifting them to the surface is pervasive in European mythology. The behaviour of the Eden killer whales in protecting whalers who had fallen overboard suggests that such mythology may have some basis in fact. The inert body of a fellow air-breather on the seafloor could quite reasonably have caused Tom to exhibit 'concern'.

The whalers of Eden may well have become Tom's surrogate family by the end of his life. Although he spent only the winter months in Eden, it is likely that his summer months, spent in unknown waters probably further south, were similarly solitary. As a naturally sociable animal, the company of his erstwhile hunting companions may well have been sufficient attraction to bring Tom back to Eden every year.

But Tom was not alone in being interested in humans. Having been scrutinised by dolphins speeding along in the bow wave of our boat, I can well believe the accounts of killer-whale researchers documenting the peculiar interest many killer whales show in humans. Captive killer whales, deprived of the company of their own kind, often form strong bonds with their trainers and human companions. Far from being a single-mindedly ferocious, predatory beast, there is something in the nature of killer whales that predisposes them to amicable interactions with humans. The scarcity of such interactions probably just reflects how rarely humans reciprocate this interest.

*O*nce there was a whale, and he ate all the fishes, until only one small clever fish was left. To keep out of danger, this small fish swam just behind the whale's right ear.

The whale said, 'I'm hungry.'

'Have you ever tasted a person?' the small clever fish whispered...

'No,' said the whale, 'what does a person taste like?'

'Very tasty,' said the small clever fish, 'tasty but knobbly. If you want to try a person, swim to latitude fifty north, longitude forty west. There on a raft, you will find a ship-wrecked sailor wearing nothing but breeches and a pair of suspenders. But be warned, this sailor is wise and resourceful.'

The whale swam as fast as he could swim until he reached latitude fifty north, longitude forty west, and there on a raft he found a ship-wrecked sailor...The whale opened his mouth wide and swallowed the sailor and the sailor's raft in one gulp. When the wise and resourceful sailor found himself inside the belly of the whale, he jumped and leapt around so much that he gave the whale hiccups.

'This sailor is so knobbly and he is jumping around so much,' said the whale, 'that he is giving me hiccups. What should I do?'

'Tell him to come out,' said the small clever fish.

'Come out,' the whale said to the sailor, but the sailor said, 'take me to my home first.'

The whale swam as fast as he could, and when he saw the shores of the sailor's home the whale rushed up onto the shore and opened his mouth wide. The sailor walked out. But while the whale was swimming, the sailor had been making a criss-cross grate, using the wood from his raft and tying it together with his suspenders. As he walked out of the whale, the sailor pulled the criss-cross grate behind him so that it stuck in the whale's throat and from that day on, the whale couldn't swallow anything except tiny, tiny Fish.

Abridged from Rudyard Kipling,
'How the whale got his throat', pp. 1–11

5

A MATTER OF TASTE

No shark will touch you with them killers there.
The killers would chop a shark to pieces.
A swordfish, you know what he's like, he wouldn't have a
chance.
An' a porpoise, he'd make a porpoise sweat he's so fast.

Percy Mumbulla, *The Whalers*, p. 20

Few predatory animals in the world can refrain from biting the hand that feeds them. Even dogs, after thousands of years of domestication and subservience to human whims, are prone to displays of ill-tempered ingratitude when their food is at stake. And when the full ferocity of instinctive blood lust is launched, there is little chance of diverting it.

The descriptions of the ferocity of the Eden killer whales attacking baleen whales calls to mind sharks in a feeding frenzy. Ripping and biting, with great chunks of blubber flying; clinging with their iron grip to the tail-flukes, lips and fins of a mighty writhing leviathan; being flung headlong through the air and leaping, 3 tonnes of furious muscle, clear of the water into physical combat with their victim—these are not conditions conducive to constraint. And yet, one of the most remarkable claims of the Eden whalers is that the killer whales not only refrained from attacking them when they were in the water, but actually protected them.

The flimsy, 9-metre open whale boats of the Eden whalers were frequently swamped, tipped and up-ended by the thrashing of a dying whale. Occasionally one of the whale's massive tail-flukes, known as the 'hand of God', would smash down across the boat, sending the men diving to the water. Whales some-times surfaced directly beneath the boat, heaving it skyward before sending it sliding into the black depths of the cold winter ocean. Injuries to whalers were common. Many whalers must have suffered the fate of Peter Lia who went missing on a dark night after the boat he was in was struck by a fin whale. It seems likely he was struck directly by the fin whale's tail.

Despite the frequency with which whalers were in the water and despite the ferocity of the killer whales' attack, no whaler was ever attacked by a killer whale. The forbearance of the killer whales towards the human whalers has been recorded through-out Eden's whaling history. In the 1840s, Oswald Brierly reported that 'although these killers will attack a whale yet they seem peaceable disposed towards mankind; for should a stove whaler get upset the killers do not seem to attempt to attack them (*Reminiscences of the sea*, MLA546: frame 10)'. As noted earlier, George Davidson claimed that after he was washed over-board Tom left the hunt to swim alongside him until his boat retrieved him. One of the Davidsons was reported as stating that 'Always, with the killers joining in this chase, a broken boat or a man in the sea, became the care of one or two killers, and until he was rescued and taken into another boat'. The previously men-tioned story of Hooky lifting a whaler bodily by the shirtfront to the surface as he sank beneath the waves seems more matter-of-fact than myth within this wealth of similar claims. Local historian H. P. Wellings wrote that 'Men have been knocked overboard during a chase but were always protected by the killers from any possible shark attack. The killers, whilst with the whale boats, never attacked any of the crew or their boats' (*Shore Whaling at Twofold Bay*, p. 5).

Aboriginal whalers in particular had great faith in the killer whales' protection. Percy Mumbulla related that 'if the whale boat was out of sight of land an' got smashed, the killers was there. They would be swimmin' round an' round, keepin' the sharks away. If them killers seen a man gettin' tired, they would swim underneath him, put a fin under his arm an' hold him up until the launch came to pick him up.' (*The Whalers*, p. 20).

Far-fetched as these stories seem, like other aspects of the Eden killer whale story they have a convincing level of internal consistency. Brierly's diaries and the accounts in the popular press were certainly written without reference to one another. Percy Mumbulla's account derives from Aboriginal oral history rather than a European oral or written account. An even more persuasive argument for close physical contact between the killer whales and the whalers is made by an almost incidental comment in the article by Hawkins and Cook describing killer-whale behaviour while being disentangled from lines: '... during the liberation process they emit a purring noise similar to the gratifying music which greets one when stroking the velvety back of the common or garden cat' ('Whaling at Eden with some "killer" yarns', p. 270). Greg McKee alerted me to the fact that this purring must actually have been the killer whale's sonar, a sound which could only be heard by humans if they were in fairly close proximity to the animals.

It is hard to believe that such voracious predators as wild killer whales would refrain from snapping up a tasty little human snack which fell in the midst of a frenzied attack on a baleen whale, particularly given the killers' legendary appetites. Like white pointer sharks, killer whales seem happy to eat anything that moves in the water. Also like white pointers, many killer whales are particularly partial to a meal of seal—something that seems readily confused with a dark-clad, splashing human. The suggestion that the killer whales actually broke off from their feeding frenzy to 'stand by' whalers overboard was, for many years, dismissed as yet another tall tale.

At the turn of the century, when these claims were first coming to public attention, the reputation of killer whales was far from benign. Indeed, the relationship between killer whales and Europeans has historically been acrimonious. Many of the early Mediterranean cultures held dolphins in high regard, but few extended this affection to killer whales. The strikingly beautiful common and striped dolphins (*Delphinus delphis* and *Stenella coeruleoalba*) were favourite subjects in the delicate murals of the Minoan civilisation. They were also great favourites of the ancient Greeks and feature in many of their myths and fables. Telemachus (the son of Odysseus) and the bard Arion were both saved by dolphins, for which latter act of mercy Poseidon created the constellation Delphinus. Dolphins had a reputation as music lovers and were therefore associated with Apollo, but as a sea power they were also the emblem of the sea god Poseidon and of Aphrodite, who was born from the sea. The sea nymphs, or Nereids, were commonly depicted as riding on dolphins, particularly Thetis (mother to Achilles). The Romans continued this fondness for dolphins, using them as emblems for both Neptune and Venus.

Unfortunately, killer whales are also very fond of dolphins, although not in a way appreciated by the ancient Greeks and Romans. Killer whales not only eat dolphins, but sometimes appear to do so in an almost gratuitously violent manner. Tom Mead, whose 'dramatised' history of Eden whaling was nonetheless based on extensive interviews with George David-son and other locals, had George talking of the Eden killer whales swimming peaceably with dolphins for some time before savagely and ruthlessly attacking the school, turning the waters red with blood and throwing the hapless dolphins in the air. Such habits were unlikely to endear them to any culture in which dolphins were messengers from the gods and saviours of shipwrecked humans. When a killer whale stranded in the harbour of Ostia near Rome, the Emperor Claudius led his Prae-torian Guard to its slaughter as a 'show for the Roman public'.

Through the Middle Ages, killer whales continued to enjoy a fearsome reputation. Aelianus described killer whales as man-eaters who 'even snatch men standing on the shore'—a behaviour they certainly apply to seals (*On the characteristics of animals*, p. 205). The Latin for whale—*orca*—came to mean a sea monster of any nature. Killer-whale bones have even been found in kitchen middens in England and their skeletons apparently accompany rock engravings in shallow caves, although little is known about how or why they came to be there.

While killer whales were encountered by Europeans in oceans the world over, it was not until humans began to explore the mysterious frozen continent of Antarctica that they encoun-tered the animal in its true abundance. As many as 100 000 killer whales cluster around the coast of the Antarctic icecap. Super-pods containing hundreds of individuals have been seen. For early explorers, killer whales were a constantly frightening feature of the icy seascape.

Explorers warily observing the killer whales in such close proximity were disconcerted to find themselves being scruti-nised by the whales in return. Killer whale vision is as acute above as below water, and they take advantage of this acuity by 'spyhopping'—raising their heads vertically out of the water to search for prey on rocks and ice floes. The Antarctic explorers were greatly discomforted by the attentions of these animals. Their fear was enhanced by the realisation that behind those piercing 'piggy eyes' lay an impressive intellect. On Thursday 5 January, 1911, the famously ill-fated Antarctic explorer Robert Scott wrote the following account in his diary:

> I was a little late on the scene this morning and thereby witnessed a most extraordinary scene. Some 6 or 7 killer whales, old and young, were skirting the fast floe edge ahead of the ship; they seemed excited and dived rapidly, almost touching the floe. As we watched, they suddenly appeared astern, raising their snouts out of the water. I had heard weird stories of these beasts, but had never associated serious danger with them. Close to the water's edge lay the wire stern rope of the ship and our two Esquimaux dogs were tethered to this. I did not think of connecting the movements of the whales with this fact, and seeing them so close I shouted to Ponting, who was standing abreast of the ship. He seized his camera and ran towards the floe edge to get a close picture of the beasts, which had momentarily disappeared. The next moment the whole floe under him and the dogs heaved up and split into fragments. One could hear the 'booming' noise as the whales rose under the ice and struck it with their backs. Whale after whale rose under the ice, setting it rocking fiercely; luckily Ponting kept his feet and was able to fly to security. By extraordinary chance also, the splits had been made around and between the dogs, so that neither of them fell into the water. Then it was clear that the whales shared our astonishment, for one after another their huge hideous heads shot vertically into the air through the cracks which they had made. As they reared them to a height of 6 or 8 feet it was possible to see their tawny head markings, their small glistening eyes, and their terrible array of teeth—by far the largest and most terrifying in the world. There cannot be any doubt that they looked up to see what had happened to Ponting and the dogs. (*Scott's Last Expedition*, pp. 94–5)

The killer whales moved on with no fatalities, but the explorers henceforth regarded them as potential man-eaters and Scott concluded that 'they are endowed with a singular intelligence, and in future we shall treat that intelligence with every respect'.

The presence of the killer whales greatly added to the stresses of an already hazardous expedition. When a party of the explorers was marooned on shifting ice, members hastened to find a path back to the main floe, only to find their route cut off by a nightmarish channel churning with swell, crushed ice and killers cruising with 'fiendish activity'. When one of their ponies fell between two floes, a party of whales immediately appeared. Apsley Cherry-Garrard, the assistant zoologist on Scott's expedition, recalled:

> 'Good God, look at the whales' said someone, and there, in a pool of water behind the floe on which we were working, lay twelve great whales in a perfect line, facing the floe. And in front of them, like the captain of a company of soldiers, was another. As we turned, they dived as one whale, led by the big fellow in front, and we certainly expected that they would attack the floe (*The Worst Journey in the World*, p. 156).

They didn't.

The tendency of killer whales to knock seals off ice floes also gave rise to a theory that they struck ocean-going yachts in an attempt to knock sailors overboard. In the early 1970s a rash of claims surfaced that killer whales had attacked and sunk yachts mid-ocean. Generally such attacks were not witnessed, but when the survivors scrambled for safety, they saw killer whales circling the scene and assumed they had been responsible for the collision. The sight of a 2-metre high dorsal fin circling the waters around a damaged boat would have been fearful for the ship-wrecked sailor, particularly given the killer whales' reputation at the time. The US Navy regarded them as one of the most dangerous species in the sea, 'a ruthless and ferocious beast' which would 'attack human beings at every opportunity'. Even more sensational were author James Clarke's claims that the killer whale was 'the biggest confirmed man-eater on earth' (quoted from a diving manual by Eric Hoyt, *Orca*, p.85) despite offering no support. Technically more accurate, although still reflecting an unwarranted bias, was diving expert Owen Lee's comment that 'there is no treatment for being eaten by the orca, except reincarnation'.

It is possible (although unlikely) that killer whales might mistake a yacht for a large whale. Killer whales do use vision to a greater extent than baleen whales but they also use sonar and it is hard to imagine how a yacht could look like a whale in ultrasound. Eric Hoyt suggested in his book *Orca* that hungry killer whales in the food-poor tropical waters where these attacks were commonly reported might launch a hasty 'surprise' attack on the supposed baleen whale, speed resulting in error. Scientists have also recently reported that killer whales tend to use ultrasound more when hunting fish, but hunt other mammals (which can often hear ultrasound) in silence. Such silent hunting would increase the chance of error.

It is also possible that affected yachts might have hit a harried baleen whale that killer whales had been pursuing. Collisions with whales are not uncommon at sea, even in the absence of killer whales. But what is most noticeable in these accounts is that the killer whales never attacked the survivors. One sailor was lashed to his trimaran overnight while three killer whales circled close enough to touch, but without attacking. Perhaps, like dolphins, they were simply curious about these unfamiliar intruders.

Despite their prodigious appetites, unparalleled ferocity as hunters and apparent taste for anything that moves, there is no documented case of a man-eating killer whale. While every other predatory species on earth (even our 'best friend' the dog) has at some time or other eaten humans, for some unknown reason, this most ferocious of predators leaves humans well alone.

There is only one documented case of a wild killer whale attacking a human. On 9 September 1972, surfer Hans Kretschmer felt something nudge his board as he lay in the surf off Monterey, California. A huge black and white creature with a massive dorsal fin grabbed his thigh but let go when Kretschmer hit it, allowing him to body surf to shore. The animal's skin felt smooth and the three deep clean gashes in Kretschmer's leg were consistent with the peg-like teeth of a killer whale. Shark skin is as rough as sandpaper, abrading the skin on contact, and shark's teeth leave a ragged wound, slicing out chunks of meat. Seals or sea-lions, with their paired ripping canines, were also unlikely to have been the culprit. Whatever the reason for this attack, it seems unlikely that the killer whale would have been deterred by a puny human fist blow. Its attack appears to have been aborted for some other unknown reason.

Yet if killer whales are averse to spilling human blood, humans certainly do not return the favour. Fishermen and whalers commonly regard killer whales as competitors for their catch, and retaliate with bullets, harpoons and spears. Perhaps the most dramatic attack against killer whales occurred in the 1950s when thousands of killer whales crowded the waters of Iceland. Icelandic fishermen complained they were reducing the fish catch and, with a level of overkill reminiscent of their fierce Viking heritage, arranged a massive retaliatory strike. The US Navy was invited to use the killer whales as target practice. Hundreds were reputedly killed with depth charges, rockets and machine guns.

Despite casualties, aggression generally does little to deter the killer whales, who rapidly learn to avoid armed boats but continue to follow unarmed boats for food. In fact, one of the most significant outcomes of the war between humans and killer whales was the realisation that there was another side to killer whales, previously unappreciated by their persecutors.

In the Johnstone Strait of British Columbia, Canada, killer whales were commonly shot at by fishermen who disliked sharing their salmon catches. Soon people noticed that, rather than fleeing the area where one of their pod had been injured or killed, the remaining killer whales would frequently try to assist their wounded colleague. Far from being cold-blooded barbarians, these whales, like many other cetaceans, clearly belonged to a complex and caring society. Compassion for a fellow whale probably led to the death of many a killer whale and it certainly led to another strange chapter in human–killer-whale interactions—the marine theme park. In the 1960s and 1970s family members who stayed to assist a wounded colleague were soon in danger of being caught themselves. Younger smaller killer whales were particularly favoured by the hunters, being easier to capture and transport. In 1965 a young male killer whale was caught in a salmon net off Namu in British Columbia. Named after the location of his capture, Namu was taken to the Seattle Aquarium. There he was befriended by the director, Ted Griffin, who astonished visitors by hand-feeding Namu salmon, swimming with him and riding on his back. People were surprised to find that killer whales, once captured, were calm, friendly animals, which interacted readily and engagingly with humans. Killer whales, like many other dolphins, were readily trained, highly intelligent and sociable, making them ideal theme park inhabitants. The suitability of a theme park for a killer whale, on the other hand, makes a less happy story. Despite some success at breeding killer whales in captivity, most captive killer whales die young and many exhibit an unusual level of aggression towards each other and occasionally to keepers, which probably stems more from boredom and captivity-induced neuroticism than malevolence.

In just a few decades, killer whales went from being 'seagoing homicidal maniacs' (*Monsters of the Sea*, p. 218) to marketable cute, cuddly toys turning tricks in swimming pools. Nowhere is this public relations reversal more obvious than in movies. In 1977, the horror flick *Orca* depicted the bereaved male killer whale as vicious and merciless, relentlessly intent upon avenging the death of his mate and offspring. By 1993, the *Free Willy* movie transformed the orca from a terrifying distributor of natural justice into an all-American kid's best friend. As academic Joseph Andriano concluded, *Free Willy* 'completely removed the monstrous, deleted the killer, from the whale'.

While their obvious intelligence and willingness to learn tricks (as will any halfway intelligent animal in a boring environment) were of significant appeal, the fact that such a dangerous animal did not eat humans suggested some kind of special relationship between species that made them irresistible. Humans have long claimed some kind of special relationship with members of the dolphin family. Dolphins do find humans interesting and they may well recognise a closer affinity between us and them than we do. Humans, in their world dominated by vision, see dolphins as sleek grey streamlined creatures bearing more resemblance to fish than to humans. Dolphins, on the other hand, literally look beneath the surface, in ultrasound, and see a skeleton and soft tissue not unlike their own, and one that's certainly very different from a fish. This could probably accounts for anecdotal evidence of dolphins interest in pregnant women. Not only can they see the mother but also the foetus, and human foetuses probably share even more structural similarities with dolphin foetuses than we do as adults.

The same argument could apply to killer whales. Do they recognise a fellow mammal in humans? They may well do, but that doesn't explain why killer whales don't eat humans. Killer whales, after all, are enthusiastic hunters of other mammals, both seals and other whales, which are far more similar to themselves than we are.

What other explanation could there be? A closer look at the list of species eaten by killer whales reveals one possibility. Not surprisingly, all the species eaten are marine species. Unlike crocodiles and sharks, there is no case of a killer whale eating a terrestrial species which has accidentally found itself in the water. If killer whales don't eat humans, we must also say that they don't eat horses or dogs or anything else terrestrial they may have come into contact with. In all the terrified accounts of killer whales mercilessly tracking Antarctic explorers there is no mention of killer whales eating one of the ponies, despite the fact that they were often lost off ice floes while killer whales were around. The killer whales that nearly knocked Ponting off the ice floe may have been more interested in the tethered dogs, but nor did they make a very concerted effort to consume them.

Perhaps the most dramatic demonstration of the killer whales' disinclination to eat humans was demonstrated by the French scientist Christophe Guinet. After several seasons of studying killer whales driving themselves bodily onto the beach to snatch elephant seal pups from the relative safety of the shore, Guinet embarked upon a remarkable test. Standing in the shallows as the great fins patrolled the waters offshore, he splashed the surface in imitation of a seal pup. Immediately, the fin of a patrolling whale oriented towards him. Accelerating, the whale began the charge which would drive it ashore to snatch its meal in an iron-fast grip. But the attack never came. Just before it reached Guinet, the whale halted, half-beached in the shallows. Displaying a formidable array of teeth in an indifferent threat gesture, the whale examined Guinet before refloating itself and resuming its patrol.

As the remarkable forbearance of killer whales has become more apparent, people have taken greater risks. Photographers filming killer whales feeding in the wild, have risked diving in with the killer whales. As they passed these strange intruders, the killer whales inspect them but otherwise show little interest. Stories like these, of which there is documentary evidence, make the claims of the Eden whalers seem quite plausible.

Perhaps killer whales just don't like eating terrestrial mammals. Captive killer whales are always fed fish—never horse meat, beef, goat or chicken, which are commonly fed to captive crocodiles and sharks. (This is partly because most captive killer whales have originated from fish-eating Icelandic or British Columbian populations.) All of the marine species upon which killer whales prey typically eat fish or crustaceans. And most things that eat fish, smell fishy themselves. In fact they generally reek of 'fish'. It might just be that the few taste buds killer whales still possess are keyed into that pungent scent. (Although their brain physiology suggests that none of the toothed whales can smell, they apparently have a highly developed sense of taste). The killer whale that let go of Hans Kretschmer in the Californian surf probably didn't do so because it recognised some kind of kindred intellect or reincarnated soul mate. Its reason may well have been far more pragmatic. To a killer whale, humans may just taste foul.

An alternative explanation is that killer whales may not like unfamiliar food. Despite having a highly eclectic palate as a species, individual groups of killer whales are highly specialised and particular about the prey they target. Humans may simply not be on their menu. But if it seems remarkable that humans should 'trust' killer whales not to eat them, consider the faith some Steller sea-lions and Dall's porpoises of British Columbia put in the dietary specificity of certain killer whales. Despite the fact that some groups of killer whales in the area actively hunt seals, the seals seem to be able to tell the difference between the seal-eaters and 'harmless' salmon-eating killer whales, consorting freely with the latter while avoiding the former like the plague. Such fine discrimination between predators by their potential prey might explain the relative frequency with which smaller dolphins are observed 'playing' around killer-whale pods as well as simply peaceably co-existing with them. If a seal or porpoise can relax in the company of killer whales, even when at risk of being eaten by their near relatives, the faith of the Eden whalers in their particular pod was probably well-placed.

Whales even penetrate into our seas. It is said that they are not seen in the Gulf of Cadiz before midwinter, but during the summer periods hide in a certain calm and spacious inlet, and take marvellous delight in breeding there; and that this is known to the killer whale, a creature that is the enemy of the other species and the appearance of which can be represented by no other description except that of an enormous mass of flesh with savage teeth. The killer whales therefore burst into their retreats and bite and mangle their calves or the females that have calved or are still in calf, and charge and pierce them like warships ramming. The whales being sluggish in bending and slow in retaliating, and burdened by their weight, and at this season also heavy with young or weakened by travail of giving birth, know only one refuge, to retreat to the deep sea and defend their safety by means of the ocean. Against this the killer whales use every effort to confront them and get in their way, and or slaughter them when cooped up in narrow straits or drive them into shallows and make them dash themselves upon rocks.

Pliny the Elder, *Natural History*, Book IX, V. 12–14

6

AN ECLECTIC PALATE

Insatiate Orque, that even at one repast,
Almost all creatures in the World would waste;
Whose greedy gorge, dish after dish doth draw,
Seeks Meat in Meat.

Josuah Sylvester, *The Complete Works*, v. I, p. 116

Killer whales will eat anything that swims—from schools of
tiny herring to gigantic blue whales, the largest living creatures
on earth. Even seabirds and sea otters have fallen victim. Their
dietary range is huge—more than 80 species have been recorded
as part of killer whale diets. Their appetites, like most carnivores, are equally vast. An adult killer whale needs to consume
about 50 to 150 kilograms of meat each day (the equivalent of
two or three seal pups). In his book, *Whales*, E. J. Slijper wrote
that 'the best illustration of the greed of a Killer Whale was
provided by the stomach of a specimen caught off one of the
Pribylov Islands (Bering Sea), which was found to contain
thirty-two full-grown seals' (p. 272). Slijper also cites Eschricht's famous paper of an autopsy of a single male killer whale
containing the remains of 13 porpoises and 14 seals. Even if (as
some claim) these victims were not eaten in their entirety, such

meals still indicate an impressive capacity for gluttony (not to mention slaughter).

The killer whale's broad diet, moreover, gives the misleading impression of it as an indiscriminate garbage guts, which eats anything that moves. But as we have seen, although killer whales as a *species* eat a broad range of prey, as *individuals* they are highly specialised. Families of killer whales in the subantarctic oceans specialise in minke whales, while those closer to the ice pack primarily eat seals. On the west coast of Vancouver, killer whales eat porpoises, sea-lions, seals and small whales. In the Johnstone Strait of northwest Canada, they exhibit such a strong preference for salmon that, after the lean winter months, the killers prefer to await the arrival of the summer salmon rather than feast on the abundance of tiny herring that mass in early spring. Yet herrings are the staple diet of killer whales in Norwegian waters. Even within the population of Johnstone Strait salmon-eaters, there is evidence of further specialisation. While the distribution of most pods can be predicted by the distribution of sock-eye and pink salmon, some pods seem to follow chum salmon instead. Similarly, within the specialised mammal-eating killer whales, different pods prefer different mammals and have different ways of capturing them. Some transient pods are found only in open water, while other specifically target the inshore waters where harbour seals haul themselves out.

In Twofold Bay, the killer whales fed on seals, occasionally dolphins, and according to whalers the 'Eden grampus' while waiting for the migrations of the great baleen whales along the east coast of Australia. The precise identity of the 'Eden grampus' is a bit of a mystery. The name 'grampus' has been applied to many different whale species across the world. Killer whales themselves were even called 'grampus' by many whalers. Former curator of the Eden Killer Whale Museum, Alex Mckenzie in *The*

Twofold Bay Story, suggested that the 'Eden grampus' refers to Risso's dolphin, whose scientific name is *Grampus griseus*. David Stead, in his *Giants and Pigmies of the Deep* also referred to the grampus as the Risso's dolphin, although this is not the species he identifies as the prey of the Eden killer whales.

Stead refers to another small whale, the little pike or piked whale, as the killer whale's preferred prey. The piked whale is more commonly known as the minke whale—a species that is indeed abundant in most of the world's oceans. The identification of the Eden grampus as a minke whale accords with a photo from about 1915, captioned 'Trying-out a Grampus: Cattle Bay, Eden', which appears to be of a minke being prepared at the works for oil extraction. Although the same size as a killer whale, an unarmed solitary minke is no match for a determined killer whale, much less a family of killers. A favourite hunting method for the killer whales was to drive small whales like minke against the shore, often causing them to strand in the shallow waters.

Trying out a Grampus

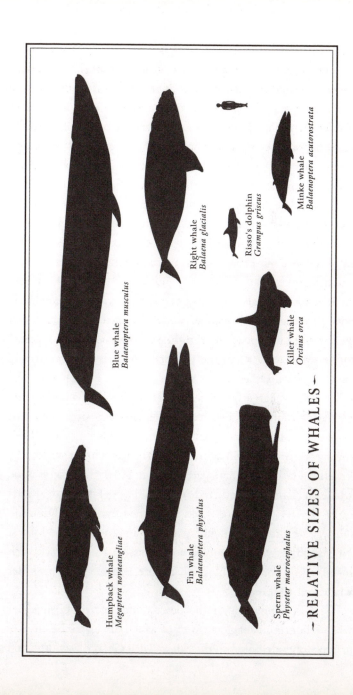

Humpback whale
Megaptera novaeangliae

Fin whale
Balaenoptera physalus

Sperm whale
Physeter macrocephalus

Blue whale
Balaenoptera musculus

Right whale
Balaena glacialis

Risso's dolphin
Grampus griseus

Killer whale
Orcinus orca

Minke whale
Balaenoptera acutorostrata

~ RELATIVE SIZES OF WHALES ~

While small whales, seals and dolphins may well have been the staple diet of the Eden killer whales for much of the year, there is no doubt that the pièce de résistance of their fare were the giants of the marine mammal world. The whales most commonly found along the eastern coast of Australia were the humpbacks. These moderately large whales leave their Antarctic feeding grounds, migrating up the Australian east coast from June to August, to give birth near the tropical islands of the Coral Sea. The tropics offer lean pickings for such large whales and they continue to feed in the bays and inlets of the Australian coast. By the time they begin their return journey to Antarctica in September with their calves in tow, the humpbacks are thin and hungry.

Right whales also frequent the coastal waters of Australia, favouring the shallow warmer waters for their winter breeding season. Both humpbacks and right whales were favoured targets of the killer whales and the whalers, being slow and relatively gentle species. Other baleen whales were also fair game for the killer whales. Large fin whales generally stuck to more oceanic migration routes but, despite their ferocity and size, could still be bailed up by the killers when they came closer to shore. Even the largest living animal in the world was fair game for the killers and their human partners—in 1910 the largest blue whale on record was cornered at Twofold Bay. This whale, measuring 97 feet (30 metres), was harpooned by Archer Davidson (George's brother), no doubt with the assistance of the killers. Nor do killer whales require human assistance to attack such massive prey. Solitary blue whales are also attacked by killer whales out to sea.

Drawing together all the descriptions of the Eden killer whale's hunting behaviour, most of which can either be directly or probably attributed to interviews with George Davidson and his fellow whalers, we can construct a likely picture of the hunting pack.

The killer whales in Eden divided into three groups during a hunt, led by Hooky, Cooper and Stranger respectively. One row stood out to sea, cutting off the escape route should the whale free itself from its pursuers. Another group of killers harried the whale from beneath, ensuring that it couldn't dive into deeper water while the third group attacked the whale. With Stranger at the head of the whale, Humpy and Charlie Adgery grappled with the whale's lips, fins and flukes, and Kinscher and Tom leapt on top of its blowhole in an effort to drown it. Big Ben and Montague assisted in the attack. Like wolves or lions, the killer whales worked together as a team to wear the victim down.

These behaviours are so characteristic of killer-whale hunting seen elsewhere that there is no reason to doubt its veracity. Similarly, although the patchy historical record does not confirm that these roles were played consistently by the same individuals, studies of killer whales elsewhere suggests that the hunting strategies observed in Twofold Bay were probably carefully orchestrated and practised by highly skilled and specialised individual hunters.

Reports of cooperative hunting in Eden by killer whales may have surprised many people at the time, when killer whales were regarded as little more than warm-blooded fish. But in fact such behaviour had long been known and reported. In 1874 Captain Charles M. Scammon described orcas attacking a whale like 'a pack of hounds holding the striken deer at bay. They cluster about the animals' head, some of their number breeching over it, while others seize it by the lips and haul the bleeding monster underwater: and when captured…they eat out its tongue' (Quoted by Hoyt in *Orca*, p.85). Modern observations also confirm that the hunting methods attributed to the Eden killer whales bear a striking resemblance to those observed in other killer whales. Scientists in Monterey Bay, California, recently observed a pod of 17 killer whales attacking a gray whale mother and calf. Small groups of the killer whales took turns harassing the whale and preventing it from escaping out to sea in much the same way as the Eden pod once behaved.

Although killer whales certainly can, and do, kill baleen whales, death may not always be their intention. Some scientists, observing killer-whale attacks, have suggested that the orcas did not appear to be determinedly attempting to kill the whales. Perhaps they were just testing individuals for weakness. Attacking baleen whales is a dangerous activity. Both Hooky and Humpy bore dorsal fin deformities, probably courtesy of the flailing tail of an enraged baleen whale. Most predators prefer to attack only weak or vulnerable individuals. Alternatively, the killer whales might simply be 'snacking' on the larger whales. Obtaining mouthfuls of flesh which, while not as satisfying as eating your fill from an entire carcass, is probably a worthwhile meal. Chunks of blubber are commonly seen floating around after killer-whale attacks giving rise to many Scandinavian names for the killers like *Spekkhogger* (or 'fat-chopper', from *spekk* meaning blubber and *hugge* meaning to chop). Norwegians also call them *vagnhund* ('to hunt like dogs' from the Old Norse *vega* 'to kill' and *hund* 'dog') or *vagnhogg*, which translates as kill and chop.

Killer whales are clearly significant predators of the large baleen whales. Between 4 and 8 per cent of bowhead whales caught in the seas north and west of Alaska bear the scars of having survived a killer-whale attack. Even higher rates of killer-whale scars have been reported for bowhead whales in Davis Strait, between Baffin Island and Greenland. A third of humpbacks caught off Newfoundland and Labrador have orca scars.

The rate of attacks must be even higher, since these figures only include whales that survived their attack and not those that succumbed. Ten dead stranded gray whales examined from 1987 to 1995 along the Chukchi Sea coast had injuries characteristic of killer-whale attacks, such as tooth rake marks. At least four of the stranded whales were missing sections of their tongue and lower lips. The Eden killer whale's preference for just the tongue and lips of the whales seems to be shared by their relatives elsewhere.

Although killer whales do hunt whales in open waters, coastal areas offer particular advantages for such predatory activities. Killer whales are masters of shallow-water hunting techniques and their nonchalant expertise in such treacherous terrain serves them well in the hunt for whales. Compared with other whales, killer whales are small, fast and agile. But they are not great divers, and a baleen whale's best defence is to head for open water and dive. Cows with calves (which cannot dive for long) cannot use this escape route and nor is it feasible in shallower waters. In 1982, three or four killer whales were observed chasing a gray whale into the shallow water of Pearl Bay, Alaska. They tore pieces from its flukes and attacked its back and sides before the gray whale 'stood-up' on its tail, perhaps in a last-ditch effort to escape its tormentors, then sank to its death. The killer whales' use of shallow water to capture other whales has been known from Roman times, and there is no reason to doubt it was used by the Eden killer whales at the turn of the nineteenth century.

Seal-hunting killer whales also take advantage of the dangerous transition zone between dry land and sea. Despite their aquatic natures, seals and penguins must return to land to breed, making the breeding-beaches of tiny oceanic islands seeth with killer-whale food—seemingly just out of reach. Where there is a will, however, killer whales will find a way.

Many killer whales patrol the shore, waiting to snatch any unwary seals and penguins as they make their way from shore to sea and back again.

Other killer whales, however, impatient with such tactics, have developed methods to snatch their prey from the apparent safety of the land and ice itself. When Antarctic explorers witnessed killer whales striking an ice floe from underneath, they were observing a behaviour regularly used to dislodge seals and penguins. Striking beneath ice floes, in unison or in sequence, cracks the ice sheet, which can be up to a metre thick, into pieces. The killer whales bump isolated ice floes to set up a regular rocking motion until the unfortunate incumbent slides into the killers' waiting jaws. On other occasions killer whales lunge out of the water at penguins and seals on the edge of thick ice. Using similar methods, killer whales have been observed taking seals from rocks in the Outer Hebrides of Scotland. But on the Indian Ocean islands of the Crozet Archipelago, some killer whales have taken such terrestrial hunting to an extreme.

With characteristic panache, the killer whales target the fat young pups on the beach itself—literally driving themselves up onto the steeply shoaling shore to snatch their prey before flinging themselves back into deeper water with a retreating wave. For some time now, French scientists led by Christophe Guinet have studied the behaviour of two pods in particular that hunt in this way. Although these killer whales will eat fish, penguins and even the odd large whale, elephant seals are their preferred staple. The whales use a variety of methods to hunt elephant seals, stalking them near river outlets and attacking them in open water, but a significant number of pups are caught by intentional stranding. Only the females and juveniles in the pods catch elephant seals in this way; the adult males are probably too large to refloat themselves from beaches once stranded.

Deliberate stranding must go against the killer whale's natural instincts. Killer whales have stranded and died along many coastlines, sometimes en masse, a fact somewhat at odds with the ease with which other killer whales negotiate shallow waters and even deliberate stranding. It may be that ocean-living killer whales, unaccustomed to the treachery of shallow waters, are more prone to fatal stranding than their more experienced coastal cousins. Some dolphin species, such as the bottlenose dolphin, *Tursiops truncatus*, live in genetically distinct 'coastal' and offshore Atlantic populations (just as the British Columbian killer whales do). Some researchers have speculated that it may be offshore dolphins straying uncharacteristically close to shore that tend to strand, rather than the near-shore populations which are presumably familiar with the peculiarities and dangers of life near the coast.

However, Ingrid Visser, whose work focuses on the coastal killer whales of New Zealand, suggests that the reverse may be true. She noticed that New Zealand killer whales have a very high rate of stranding and that live stranding (as opposed to dead animals washing ashore) tends to occur on the sandy flats near where the killer whales feed in shallow water. Visser argues that it is precisely their risky shallow water hunting strategies that make these killer whales vulnerable to stranding.

Intentional stranding is a highly skilled manoeuvre and it took the female killers of the Crozet Archipelago pod studied by Guinet since the late 1980s many years to teach their daughters Zoe and Mae how to perfect this art. Even this training, however, doesn't provide a perfect safeguard. When young, Mae stranded too far up the beach and was unable to refloat herself. She was found by the researchers calling to her mother who was patrolling offshore. With considerable effort, the researchers managed to turn the half-tonne whale towards the sea and, on each breaking wave, were able to assist Mae to fling herself back into deeper water.

Despite the difficulty and dangers of the stranding hunting method, the whales of the Crozet Islands are not the only animals to employ this technique. Eight thousand kilometres away, on the edge of the southern Atlantic Ocean, killer whales use a similar beaching technique to catch sea-lions on steeply sloping Patagonian beaches. These steeper beaches allow the whales to flick themselves back into the water more readily than the Crozet killer whales, and even the large males, which weigh in excess of 6 tonnes have become proficient at stranding. Two brothers, who have been known to successive scientists as Big Mel and Bernd, regularly snatch meals from these Argentinian sea-lion colonies. Another well-studied pair of killer whales which use similar beaching strategies, Blanche and Serge, exemplify the way in which these animals commonly hunt in pairs, although groups of up to twelve are also seen.

The skills and perils of shallow-water hunting need not always be as dramatic as bursting through the surf, jaws agape, on an unsuspecting seal pup. In New Zealand, killer whales tackle dangerous prey just as skilfully in the shallow mudflats of the North Island. Here killer whales are often seen diving in water so shallow their tails are sometimes left thrashing in the air while their snouts surface covered in sticky mud. Their quarry are large bottom-dwelling stingrays and eagle rays, all of which grow from 1 to 2 metres in wingspan. All of these rays have venomous spikes on their tails, which have been known to inflict fatal injuries on bottlenose dolphins. A dead killer whale was also found with stingray barbs embedded in its spine and neck muscles. One killer whale was observed avoiding this hazard by pinning a ray to the bottom while a second whale grabbed it by the tail thereby immobilising these weapons. More commonly, the killer whales squirt a stream of air bubbles into the mud, perhaps in an effort to startle or dislodge their prey.

Killer whales are certainly not averse to tackling well-armed or dangerous prey—as evidenced by their taste for the giant whales, as well as aggressive sea-lions and venomous stingrays. They have an almost blasé attitude towards tackling animals which are themselves predators. Percy Mumbulla commented that the 'killers would chop a shark to pieces [and] a swordfish, you know what he's like, a swordfish wouldn't have a chance'. Killer whales also represent one of the few threats to perhaps the most formidably armed ocean predator of all—a predator whose very name makes a surfer's blood run cold—the great white shark.

White pointers evoke our deepest fears, perhaps because of the dispassionate manner in which they will dispatch a human as readily as any other piece of meat. They are large sharks, up to 5 or 6 metres long, and equipped with jagged rows of serrated teeth and an almost mindless determination to apply them to anything even vaguely edible. For millions of years, this predator scoured the cold oceans of the earth virtually unchanged and, perhaps, unchallenged—until killer whales appeared. At more than twice the weight, just as fierce and with considerably more brainpower, the killer whale is more than a match for the well-armed white pointer.

In 1997 white-pointer researchers in California saw a killer whale striking a 3- or 4-metre long shark under water, apparently killing it with a single blow. The killer whale and its companion carried the dead shark around like a trophy before finally eating its liver. Killer whales have also been observed feeding on sharks in Costa Rica and New Zealand, suggesting that such shark killing is not unusual. The Californian researchers noted that, for the next two months, the temporary presence of killer whales in the area was accompanied by an unusual absence of white sharks.

Perhaps killer whales regard white pointers as competitors for the same food—particularly seals—in much the same way as lions regard cheetahs on the African plains. And just as lions will take any opportunity to kill young cheetahs (being the major cause of death of cheetah cubs), it seems that killer whales may do the same to white pointers. Perhaps there is, after all, some substance to the popular beachside belief that dolphins keep sharks away, if only for the largest 'dolphin', the killer whale. Such observations certainly add weight to the Eden whalers' claims that no sharks would come near their whale catches while the killer whales were about, despite their abundance at other times.

Hunting in pairs or small groups undoubtedly makes tackling potentially dangerous prey, from sharks to large whales, much safer. Among the mammal-eating killer whales of British Columbia (known to scientists as 'transients') it seems to maximise the chances of locating and capturing prey like seals, sea-lions and porpoises. But such social foraging also offers other advantages—the opportunity for genuinely cooperative hunting. For example, as well as their dramatic stranding strategy, the Patagonian killer whales also hunt sea-lions and elephant seals in a group, with several animals flanking the intended prey to keep it from escaping. Even more sophisticated are the cooperative hunting methods some killer whales employ to catch fish.

Not all fish-hunting requires cooperation. Killer whales in British Columbia hunt salmon (*Oncorhynchus* spp.) in groups but, apart from helping each other to find food, they capture their prey individually. Salmon are large enough to be pursued individually by the whales—each fish constituting a reasonably sized snack. Much more difficult to make a meal of are the tiny north Atlantic herring, which are the favoured food of Norwegian killer whales. If the Norwegian killer whales tried to hunt down each herring individually, the effort would hardly be worth the resulting catch.

However, Norwegian killer whales have developed a remarkable method of concentrating and killing their prey, ensuring that when they stop to eat, they have a meal worth waiting for.

The researchers who study these killer whales refer to their method as 'carousel feeding', in reference to the relaxed circling used by the killer whales to round up the herring. Killer whales initially select herring schools most vulnerable to attack. Smaller schools seem less able to perform the range of dramatic anti-predator manoeuvres (such as splitting, diving and forming various shapes to confound the killer whales' herding efforts). Round, dense schools are also favoured by the killer whales. Lazily circling so as not to disturb the herring, the killer whales break up and position themselves around the school. Communicating in sonar, which seems not to affect the fish, the killer whales suddenly twist their dark bodies so as to 'flash' the herring with their white underbellies. This flash alarms the herring, which dart away, only to be met by another flash on the opposite side. Each time the school streams in one direction it is startled into the other direction, condensing into an ever-tightening cloud of frightened fish. In a matter of minutes, the school condenses, and the killers in deepest water begin to flash underneath the herring, gently bringing them up to the surface in a seething mass.

The temptation must be for one whale to dive into the herring, snapping up as many as possible before they disperse. But the Norwegian killer whales would pass a simple intelligence test that many humans fail—the ability to renounce a short-term individual reward in order to achieve a greater reward for the group in the future. Instead of diving in and grabbing a single mouthful, the innermost whales deliver powerful slapping blows to the herring schools with their tails, while the outer whales continue to circle and prevent the herring from dispersing. Only once the water is filled with the stunned and broken bodies of thousands of tiny herring do the killer whales break from their slaughter and all begin to feed, delicately snaffling up each tiny herring one at a time.

The level of coordination required by the herring hunters, like that of the whale hunters, is clearly high and could only be conducted by an animal with a high degree of socialisation. Likewise, the degree of socialisation required for the seal-hunting killer whales of the Crozet Islands is also high—not so much because they need assistance to actually kill their prey, but because they can only acquire their sophisticated hunting techniques through a long period of active learning. Unlike many other solitary predators, killer whales are not all born successful hunters, they must learn their trade. And some individual killer whales learn their trade better than others—becoming the primary hunter for their family and ensuring that the rest of their pod is kept in food. Even among other predators, such as the big cats which also learn how to hunt, killer whales are unusual in the length of their period of development, the diversity of hunting methods employed by different family groups and the deliberate 'training' of young animals to hunt. The readiness with which wild killer whales learn such a striking diversity of complex and highly skilled hunting behaviours makes 'learning' to cooperate with human whalers in Eden almost mundanely simple by comparison. It took the whales fewer than fifteen years to learn to work with the European whale boats (from the time of the first whaling operation in 1828 to Oswald Brierly's descriptions of the hunt in 1843). Perhaps the remarkable feature of the Eden killer whales is not so much that they learnt to cooperate with humans, but that this is the only instance where humans learnt to cooperate with them.

*A*cross the Bay at Edrom
   Where the Kiah River flows
And meets the Blue Pacific on its bay.
The smoke drifts up unheeded
Where the melting pots are needed
For the whaling season now is underway.
And you can see the big sweeps swinging
And the whale boats' crews are singing
George Davidson in the bow as plain as day.
And the wily killers playing
Near the longboats sometimes straying
As they lead them to the whales in Twofold Bay.
And you can see the water flying
And the mighty whale is dying
For the hand harpoon has met its mark with skill.
And the longboats lie abreast
Whilst the crew stand off and rest
And the hungry killers move in for the kill.
With six strong men in his longboat crew
With hand harpoon and lance,
He fought his quarry face to face
and gave the beasts a chance
With aircraft flying overhead, chaser boat and gun
Its no longer now to kill the whale,
But murder, murder everyone
And the old whale knows as she journeys south
In a world that's filled with greed
Her chance to meet the mating bull is very slim indeed
And she sounds, she flicks her tail
In a sheet of flying spray
And she longs again to do battle with
"The Man from Twofold Bay".

From Kevin Warren, 'Davidson the Whaler' quoted in
René Davidson, *Whalemen of Twofold Bay*, p. 95.

7

A FAMILY AFFAIR

All set with iron teeth in ranges twain;
That terrified his foes and armed him,
Appearing like the mouth of Orcus grisely grim.
Edmund Spenser, *The Faerie Queene*, VI, xii, 26

The predatory prowess of the killer whales undoubtedly stems from their strong family bonds. All killer-whale hunting strategies are highly individual, often complicated, and almost certainly learnt, after long apprenticeships, from older animals in their pods. Whether they capture their prey alone or in a group, killer whales require assistance from each other, either to physically hold their prey or to learn how to catch it.

The complex and risky intentional stranding of the Crozet Island killer whales, for example, is mastered only after a long period of training. Infant killer whales suckle for the first year of their life and remain very close to their mother for the first two years. It is not until the third year that young killer whales begin the serious business of learning how to hunt, and even then it takes several years before they are sufficiently independent for their mother to have a second infant. Mae and Zoe and other young killer whales began stranding alongside the adult females within the first few years of life. Generally they stranded with their mothers, although one calf whose mother stranded infrequently typically practised stranding with the other more skilled females of the pod. By the time they were four or five years old, the young killer whales had begun stranding on their own, but it was not until they were five or six that they first caught elephant seal pups by this method. Even then they still required the assistance and encouragement of an adult female to help them return to the water with their prey. The apprenticeship for this method of hunting is long and relies on the strong life-long social bonds so characteristic of killer-whale society.

Group hunting also requires a strong social unit. Killer whales are like many terrestrial predators that hunt in groups to round up small prey or bring down prey larger than a single hunter could manage. Like Nile River crocodiles and Amazonian giant otters, pelicans and dolphins, families of killer whales circle, drive, corral and ensnare schooling fish in natural traps against rocks and inlets. And like wolves, lions and hunting dogs, individual killer whales within the pod take on different tasks when hunting large whales—some harrying from behind, others clinging to limbs and appendages, another flinging itself on top of their hapless victim. The reported behaviour of the killer whales in Eden bore the hallmarks of both coordinated action and carefully acquired skills which have been observed in other killer whales. The distinct structure of the hunting party in dividing into three groups, each of which took up different stations, suggests a division of labour borne of experience in working together. The individual specialities of particular killer whales suggests that their skills as hunters were both learnt and valued—they were not roles that could be fulfilled by just anyone, some killer whales were better at certain tasks than others.

In contrast to the well-known cooperative hunters—wolves, lions and killer whales—most predators tend to be individual hunters, defending their food against others, as a domestic kitten or puppy does, growling fiercely over its prized morsel of meat. I've always had a special interest in what makes animals social when the vast majority of species are not. Perhaps because we are social animals, the origin of sociality is particularly intriguing, but the answers are not always as obvious as they first appear. Most species seem to prefer to keep as many resources to themselves as they can. Cooperative hunting arises only under quite unusual circumstances, typically when defending one's own food is either not possible or not worthwhile. Small reliable sources of food, like a barn full of mice, for example, make it both possible and worthwhile for a domestic cat to keep rivals away. A lion, however, cannot prevent other animals from preying on migratory herds of antelope scattered across large areas. Schools of fish or migrating whales are also almost impossible to defend.

Nor is there any point in defending such abundant food. When thousands of migrating wildebeest flood the African plains, there is enough food available to feed many predators for a short time. Fish provide a similar short-term food boom. Fish-eaters cannot predict when or where their food will be at any one time, but when they find a school there is more food available than any one individual can possibly eat in the short time before the fish disperse or disappear. Large meals also reduce food competition between individuals. A single wildebeest can feed a number of lions or hunting dogs. A baleen whale is far too much for even a pod of killer whales to consume in one sitting (which is perhaps why the Eden killers only ate the tongue and lips). With minimal competition for food, cooperation can potentially flourish amongst the most surprising of hunters—from crocodiles to killer whales.

Although reduced food competition is a prerequisite for cooperative hunting, it doesn't automatically result in cooperation. Species like seabirds rarely compete for food at sea (despite their vociferous defence of limited food like hot chips or bread), but don't actively cooperate to catch fish. In fact, cooperation often involves trading short-term individual losses for longer-term gains and occurs in species that have already sacrificed some individual benefits in order to live socially. Cooperative hunting provides benefits that help outweigh some of the costs of social living, including competition for food, mates and shelter and increased risk of diseases and parasitism. Animals live socially for many reasons but rarely do so in order to hunt cooperatively. Cooperative hunting tends to evolve after a species has adopted a social lifestyle—it is an added advantage rather than a primary motivation.

The best studied example of this evolutionary trend comes from lions. Lionesses were once thought to live in groups in order to hunt together. But, despite being able to tackle larger prey through group efforts, lionesses in groups do not necessarily eat more than individuals hunting alone. Lionesses actually seem to form 'sisterhoods' to defend their young against infanticidal male lions who try to kill all the young of the previous incumbent male, bringing the females back into season. 'Sisters-in-arms' can defend their young where an individual cannot, forcing a new male to wait for the cubs to mature before taking his turn as sire. Protecting their young is so valuable to the lionesses that it outweighs the costs of social living, like sharing food. They maximise this benefit (and minimise the cost) by hunting cooperatively and living in closely related groups (thereby helping their relatives rather than unrelated strangers).

Like lions, killer whales also give the impression that they live in groups in order to hunt large prey cooperatively.

Whale-hunting killer whales are commonly found in groups of more than ten, while killer whales hunting seals (which can be caught by one individual) occur in smaller groups of three or four. Many researchers assume that prey size (or the difficulty of catching it) dictates the size of killer-whale pods.

This assumption cannot be correct. The salmon-eating killer whales of British Columbia are amongst the most social of all (commonly foraging in pods of more than ten), and yet they use little or no cooperation to catch their prey. Rather than joining forces in order to catch large or difficult prey, killer whales live in larger groups wherever there is enough food to reduce food competition—irrespective of whether that food is caught individually or cooperatively. Cooperative hunting may be one of the benefits of killer-whale society, but it doesn't appear to be the driving force that keeps killer whales together. In fact, rather than coming together in order to hunt, I think large killer whale families split up when food is scarce. When food is abundant, killer whales can indulge their natural proclivity for a social lifestyle.

Most of our knowledge about killer-whale family life comes from the killer whales of British Columbia, particularly the relatively stable in-shore salmon-feeding pods known to scientists as 'residents' of the Johnstone Strait. For the last 30 years, these 200-odd animals have been studied intensively by successive generations of students. Each individual in the population has been named and indentified, its family and personal history noted and recorded.

The resident killer whales studied in British Columbia can be divided into four clans, all of which share common calls, which they use to communicate with one another and identify clan members. Each of these clans divide into between two and ten pods, which comprise from ten to 50 animals. These pods then break down into sub-pods, which generally consist of stable matrilineal units which can include a mother, her children, grandchildren and even great-grandchildren. Pods generally only divide upon the death of their matriarch, although even then smaller families of sisters remain together after the death of their mother. These sub-pods are the most cohesive social units with both male and female calves staying in their mother's pod for their entire lives. Each sub-pod shares calls with all the other pods of its clan, but also exhibits distinctive variations of its own.

Females play a leading social role in the sub-pods. Adult males tend to travel at a distance from the centre of the pod, and often forage in the poorer deeper waters. Females and calves forage in a closer group and take up the best foraging grounds. They are led by the matriarch, the oldest female, who usually comes up to breathe first. When two pods meet, the matriarchs often swim side by side, linking the two pods together. During the summer months, when salmon are abundant, even larger groups of killer whales are often found together. These salmon-hunting whales are not cooperative in their hunting techniques and so these large groups do not come together to assist with food capture. However, the abundance of food available during the salmon-run season in this area enables the killer whales to aggregate without competing with one another for food. Such relaxed socialisation is a luxury not available to all killer whales.

Another type of killer whale also inhabits the waters of British Columbia. 'Transient' killer whales were so dubbed by researchers in the 1970s because they seemed to be less permanent than the salmon-feeding resident pods. Further research over the years since then, however, has revealed that the seal-feeding transient killer whales are just as much resident in their larger home ranges as the salmon-feeders are to theirs. The names

TYPICAL KILLER WHALE POD STRUCTURE

This sub-pod of British Columbian resident killer whales illustrates the strong family ties underlying the structure of killer whale pods. The current matriarch of this pod is known as C6 who travels with her presumed uncle C3, four surviving offspring and two grandchildren. Solid lines indicate known relationships between killer whales (confirmed by observations of nursing mothers and young) while the dashed lines are the assumed relationships based on the association patterns and age class of the whales when first identified (adapted from *Killer Whales: The natural history and genealogy*).

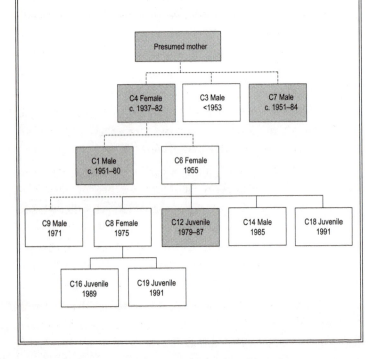

of the two populations, however, have stuck. Transient killer whales have a slightly different social structure, which probably reflects their preference for eating smaller marine mammals such as porpoises and seals. Although the pods are still matrilineal, they are much smaller—often consisting of just three or four animals. This group size is similar to the seal-hunting killer whales in Argentina, which typically feed in groups of two or three animals although pods of four or five are common.

The bond between the mother and her adult offspring remains strong between the transients, just as it is among residents. But, unlike the residents, young transient killer whales sometimes leave their mother's pod, particularly when there are more than three animals already in the pod. Scientists have calculated that pods of three killer whales receive the most food per whale. Three killer whales probably maximises the whales' success at detecting and hunting their favourite prey of harbor seals. Additional pod members do not increase the number of seals caught, but decrease the amount of food shared out to each whale. Thus, by the time a mother has had her third calf, the pressure is on for one of her older offspring to leave the pod.

Among transient killer whales, the oldest adult son tends to stay with his mother, but younger sons may be forced to leave the pod and become roaming solitary males. Unlike other dolphins (and some other killer whales), solitary adult transient males rarely associate with one another. Females, on the other hand, are more determinedly social. When forced to disperse from their mother's pod, solitary females temporarily join up with other pods, if only for a few days. Females are never seen alone. Such temporary associations are, however, unlikely to fulfil the female whales' need for a social network in which to breed. After the disappearance of a sibling, one adult female who had dispersed from her mother's pod two years earlier promptly returned to the family fold. Transient killer whales would probably prefer to stay with their mother's pod, just like residents, if there was enough food to go around. When they are not hunting, the small pods of transient killer whales are just as sociable as residents, socialising together in large leisurely groups—it is only when feeding that the smaller units are maintained.

The Crozet seal-hunting killer whales form similar-sized groups, however scientists have observed an interesting behaviour in these killer whales that may explain variations in family-group sizes. Although killer whales are generally very quiet when hunting whales and large elephant seals, once they have caught their prey the whales emit long-distance contact calls. Members of their own pod and other neighbouring pods soon arrive from kilometres away. Christophe Guinet has suggested that such calling behaviour may allow larger groups to form when there is more prey, or larger prey, available. This tendency to form large groups when there is an abundance of prey is well-illustrated by the Norwegian herring-hunters, whose pods average about fifteen animals, allowing them to exploit sophisticated cooperative hunting methods.

The ultimate superabundance of food, however, comes from hunting large whales. On the whole we know little about the whale-hunting killer whales, most of which spend their lives dispersed across large areas of ocean in search of their equally wide-ranging prey. Scientific observations of these animals tend to be confined to opportunistic observations of whale hunts. The individual killer whales themselves, their relationships and social lives remain a mystery.

From lucky observations of whale hunts it is clear that the larger the species of whale being hunted, the larger the pod of killer whales involved in its capture. Attacks on small whales, such as minke, are typically attended by smaller pods of three to seven killer whales. Gray whales are commonly targeted by groups of three to ten animals. Although gray whales can grow up to 15 metres long, typically the animals attacked by killer whales are less than 10 metres. A pod of fifteen killer whales was seen attacking a Bryde's whale (which grow up to 14 metres long) and calf in California. Humpbacks, which reach lengths of up to 16 metres, have been seen under attack from pods of ten or twelve animals, while eleven killer whales were seen attacking blue whales off the South Australian coast. Groups of sperm whales, whose aggressive natures and size (up to 18 metres long) are the stuff of legend, have been seen attacked by pods of 15 to 25 killer whales.

Similarly the Eden killer whales, living in a pod of at least 25 to 30 animals, typically hunted larger baleen whales of 15 to 30 metres. This large pod probably broke up into its three smaller constituent sub-pods of seven or so animals to pursue smaller whales and porpoises. These smaller family groups lead by Hooky, Stranger and Cooper may well have been the usual social unit for the Eden killer whales in summer. It may just have been in Twofold Bay, at the height of the baleen whale migration, that the killers would all join forces.

Despite the increasing size of killer whale pods associated with larger prey, the number of individual killer whales actually attacking the whales remains small—between one and seven animals. Thus, the large numbers of killer whales seen during attacks on large whales are not there because it takes more animals to capture them (although they may well attack in relay), but because there is more food to share at the end of the hunt.

If killer whales sometimes hunt cooperatively because they are social animals (rather than vice versa) it begs the question of what makes them social in the first place. With the vast majority of species on earth opting for a solitary existence, social animals

EDEN POD STRUCTURE

Hooky's sub-pod is the only group whose membership is known with any certainty. The remaining animals must have belonged to either Stranger's or Cooper's families, but we don't know which individuals belonged in which sub-pod.

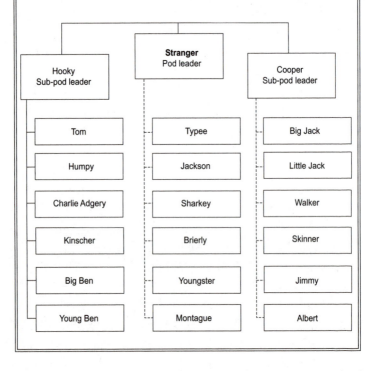

are the exception. As noted before, social animals run the risk of diseases and parasites as well as increased competition for mates, food and shelter. To understand what makes killer whales social, we must take a closer look at what their life-history traits tell us about their social life.

The Haida Gwaii people of Queen Charlotte Island, Canada, regard the distinctive dorsal fin of the killer whale as a doorway to a watery underworld. For scientists, the killer whale's dorsal fin is also a doorway—into the watery world of the killer whales themselves. The distinctive shapes and patterns of the dorsal fin and patch allow individual identification, which is critical to the scientific study of populations. The nature of the dorsal fin itself tells us a great deal about the probable social lives of these somewhat mysterious creatures.

It was for this impressive feature (standing up to 2 metres tall on some males) that many northern Europeans named the killer whales swordwhales or swordfish—*Schwertwal* in German, *zwaardwalvis* in Dutch and *sverðfiskur* in Icelandic. In Norway, the killer whales are also named after their pointed dorsal fins—*staurkval* or *staurhyrning* (from *staur*, meaning pole or fencepost, *kval* meaning whale and *hyrning* meaning 'having a horn'). Modern Icelanders use the same word base in *háhyrningur* to refer to killer whales. The Latin *orca* for whale may ultimately have derived from the old Greek *oruga*, the accusative of *orux* meaning pick axe or spike, which has more obviously given rise to the name of the scimitar-horned oryx. Such origins are also reflected in unlikely stories of killer whales using their dorsal fins to attack large whales.

Far from being the aggressive weapon implied by these human aspersions, the killer whales' dorsal fin probably prevents aggression. Although both sexes have moderately high dorsal fins in comparison with other whale species, it is only in the males that the fin becomes attenuated and extremely high. As the males reach sexual maturity at six to ten years, their fins grow up to 2 metres high, in a thin triangular shape. While food, shelter and sex are all things animals compete for, only competition for sex results in one sex being a dramatically different size, shape or colour from the other. Not only do male killer whales possess out-sized appendages, they are also larger in body size, which also suggests that the males are competing with one another for mating opportunities. The 'sexual play', including mock mounting and rubbing, frequently observed between male killer whales in the wild might well be part of a range of behaviours that determine dominance or mating precedence. Like the peacock's tail and the lion's mane, male killer whales presumably use their dorsal fins to intimidate their rivals and impress females, rather than for fighting with each other. Little is known about the details of killer-whale social etiquette, but it seems likely that females choose their mates on features such as fin size, while males probably size up each other's fins to determine who will have access to a receptive female rather than risk a costly fight. Male killer whales are often observed, for example, floating head to tail, in sequence holding their enlarged pectoral fins out of the water.

Only further observational studies and genetic paternity studies will cast light upon these mysteries of killer whale behaviour. For the time being, the killer whale's physical features suggest a polygynous species, in which one male mates with more than one female and, importantly, in which reproductive success among males is unevenly distributed—some males will have all the luck and others have very little at all.

The flip side of competition between males is that the females are probably very particular in choosing their mates. With many males competing for her attention, it behoves a female killer whale to choose the father of her offspring wisely. Young killer whales take up a huge amount of a mother's time and energy. Gestation takes about sixteen months. Infant mortality is very high among killer whales. In British Columbia, up to 40 per cent of resident calves die in their first year of life.

Thus, although females are capable of breeding every three years, in practice an adult female tends to produce only four or five young that survive to maturity in the course of her life. Even once they pass into the safer 'juvenile' years, young whales remain highly dependent upon their mothers.

AVERAGE LIFE HISTORY TRAITS OF KILLER WHALES AND HUMANS

Trait	Killer whales	Humans
Weight at birth	40% adult weight	0.05% adult weight
Length at birth	2.1–2.7 m (34% adult size)	0.5 m (25% adult size)
Growth rate	31 cm/year	13 cm/year
Male growth spurt	5–6 years old	15–18 years old
Male age sexual maturity	10 years old	15 years old
Female sexual maturity	15 years old (5 m)	15 years old
Age at reproduction	10 years old	20 years old
Gestation	12 months	9 months
Average time between infants	4 years	3 years
Average number of adult offspring	4–5	2–8
Age females cease reproduction	40 years old	45 years old
Years as 'grandmother'	20–30 years	30–40 years
Average female lifespan	60–80 years old	80 years old
Average male lifespan	40–50 years old	70 years old

Like humans, killer whales probably need a great deal of social support in order to successfully raise their young and, like humans, they have evolved an unusual method for ensuring that such support is available. While most animals reproduce throughout their lives, female killer whales cease reproduction up to 40 years (and commonly ten to fifteen years) before they die. This post-reproductive life probably allows the 'grand-mother' killer whale not only to raise the last of her own offspring to maturity, but also to see each of her daughters established with their own families before she dies and leaves them to their own devices.

The similarities between the life parameters of killer whales and those of humans are striking and probably result from similar evolutionary pressures. Both species are long-lived. Both species reach sexual maturity relatively late in life and have a long period of immaturity. And the females of both species cease reproduction in the mid to latter years of their lives. Even the tendency for females to live longer than males (which is quite pronounced in killer whales) is echoed to a lesser degree in humans. In both species, these life history parameters are probably the result of the high level of learning required by their young before they can survive (and breed) on their own. In killer whales, it is clearly their unusual learnt hunting behaviours that need to be acquired by the young and the role of mothers in passing on these skills, as has been clearly demonstrated by the Crozet killer whales.

The extreme diversity and complexity of killer whale hunting strategies often mean that while some killer whales may be highly successful and competent predators, many pod members will be less successful. Living in groups ensures that each family has a moderate 'skill pool' upon which to draw, increasing the chances of at least one pod member being a highly successful hunter. Group living ensures that such experience and skill is passed on with certainty to the next generation, rather than relying on hit and miss genetic inheritance of behavioural traits. There is even some physical evidence that this cultural transmission of information has been a significant feature in shaping the evolution of killer whales.

Killer-whale genetic variation is remarkably low, particularly mitochondrial variation. Mitochondria are the energy power-houses that are found in the cytoplasm of every cell in our body. They have their own genetic material separate from the genes in the cell nucleus that we inherit from both our mother and father. Mitochondria, by contrast, are only inherited from the mother, passing on to each child from the cytoplasm of the egg, which ultimately gives rise to all other cells.

Whale researcher, Hal Whitehead has suggested that the unusually low level of mitochondrial genetic variation in killer whales is the result of strong selection for cultural, rather than biological, traits in matrilineal groups. For example, when a mother develops a particularly successful hunting strategy (such as intentional stranding or carousel feeding), she teaches it to her offspring, giving them a significant survival advantage over other killer whales. These behaviours are inherited, not through genes directly, but through culture. Such a familial advantage results in a large number of killer whales being descended from a single successful mother, dramatically decreasing the genetic diversity of the whole population. A similar pattern has been observed in the inheritance of dialects amongst British Columbian resident killer whales. While each pod has its own unique vocalisations, they also share features of a common dialect with other pods, suggesting that they are all originally descended from the one family. All of the pods belong to one of four dialect clans, giving the impression that all the resident killer whales in the Johnstone Strait are ultimately descended from just four females.

Understanding the principle which underlies fluctuations in killer whale social groups might explain the changes to pod structure in Eden. Some reports noted that in the early heyday of whale hunting, hundreds of killer whales were observed in the area. By 1878 (when shore-based whaling had all but wound up in most areas), only 27 killer whales participated in the hunt. This somewhat large pod may have been sustained on relatively low numbers of baleen whales because their success at hunting (aided by humans) was unusually high. Even this assistance, however, may not have been enough to maintain such a large pod. As the number of baleen whales continued to decline, under constant assault from offshore whalers, the size of the hunting pod in Twofold Bay also declined, until finally just two or three of the older males were left. With the last baleen whale being captured in 1926, it is less surprising that most of the killers left and more of a mystery that these last few elderly animals continued to make their annual pilgrimage back to Twofold Bay. Tom, in particular, seems to have shown a loyalty to Eden—which could hardly be explained by the presence of the odd dolphin, seal, grampus or even fish from George Davidson. Something else brought Tom back to the old whaling station every year. Social bonding, companionship, familiarity—if such a thing is possible between killer whale and human—it might even be called friendship.

*W*hen the Perano whalers [of Cook Strait in New Zealand] killed the whale off Jordy Bay they found to their amazement, deeply embedded in its body, an old toggle-harpoon.

What mystified them was that this harpoon was not of a kind that any of the Peranos or their veteran employees had ever set eyes on. They could clearly identify markings on the old-fashioned hand-thrown harpoon. They could discern the initials 'BB' engraved on one side and 'HO', a foundry mark of some kind, on the other...

By coincidence, Joe Perano (senior) was about to make one of his war-time trips to Eden, southern NSW, where he was planning and supervising the setting up of the proposed whaling factory. For some reason known only to himself, he took a careful mental note of the initials and shape of the old harpoon. It so happened that a week or two later Joe Perano found himself sitting in George Davidson's office in the little port of Eden.

A naturally observant man, Joe Perano looked about the office. His gaze settled on a beautifully hand-carved miniature whale-boat that would instantly endear itself to the heart of any true-blooded whaler. Suddenly, Joe Perano became excited. His heart pumped faster—it could not be!...It was an exact replica of one of George Davidson's own former whale-boats. But what had excited Joe Perano so much was an incredibly minute detail. Within the miniature row whale-boat of George Davidson's lay a tiny harpoon—again complete in detail...

...Joe was quick to notice that the miniature hand-carved harpoon was, indeed, a toggle harpoon of a rather distinctive design...Joe Perano reached for his notebook. He produced for old George Davidson the initials 'HO' and 'BB', plus a simple but clear sketch of the old harpoon ...'That is my harpoon' shouted George Davidson...He realised that this harpoon was undoubtedly the one that had been thrust into a bull right whale off Twofold Bay, NSW, 17 years earlier.

D. Grady, *The Perano Whalers of Cook Strait*, p. 97

8

PARTNERS IN CRIME

Of this whale little is precisely known to the Nantucketer, and nothing at all to the professed naturalist. From what I have seen of him at a distance, I should say he was about the bigness of a grampus. He is very savage—a sort of Feegee fish. He sometimes takes the great Folio whales by the lip, and hangs there like a leech, till the mighty brute is worried to death. The Killer is never hunted. I have never heard what sort of oil he has. Exception might be taken to the name bestowed upon this whale, on the grounds of its indistinctness. For we are all killers on land and on sea: Bonapartes and sharks included.'

Herman Melville, *Moby Dick*, p. 173

The relationship between George Davidson and Tom in later years may have resembled the kind of personal bond that sometimes develops between individual humans and individual wild animals. But the foundation of this friendship was laid in a business partnership between the 30 or more killer whales that frequented Twofold Bay and the generations of whalers who made a living there. Such collaborations between different species are rare, and there are many possible ways the association between the Eden killer whales and whalers might have developed.

An initial step might have been one of mutual exploitation, which soon grew into cooperation as each party came to realise the benefits of the other. Killer-whale attacks could well have alerted early European whalers to the presence of a whale, and whalers would have had no compunction about taking advantage of the whale's vulnerablity under the circumstances. The killer whales, however, may have had a more ambivalent attitude towards these early whalers. The killer whales had probably been using Twofold Bay as a natural trap for migrating baleen whales long before the whalers arrived. Suddenly, the whales they pursued into the Bay were harpooned by men in boats who tried to drive the killer whales off their quarry. After a long, tiring and dangerous chase, the killer whales were often forced to relinquish their hard-earned catch to the whalers who towed it inshore without so much as a thank you.

Humans have a long history of commandeering the hunting capacities of other more talented predatory species. Dogs, for example, have been bred for hunting for nearly 3000 years, as evidenced by a hunting party of mastiffs painted on the palace wall of the great Assyrian king, Ashurbanipal, in 645 BC. Similarly cormorants fitted with special collars and floats have been used for fishing in China and Japan since at least 317 BC. This method of fishing spread to Europe as a fad for the wealthy, with King Louis XV of France keeping cormorants in the sixteenth century, and King James I of England appointing a Master of Cormorants in 1611. The fashion of cormorant fishing was revived in eighteenth century Europe by a certain falconer, Captain Salvin. Cormorant fishing is still practiced today in China and Pakistan, as is fishing with otters. Ferrets are yet another species with a long hunting association with humans. Ferrets used to bolt rabbits from warrens in modern Europe and beyond, just as they were in Roman times.

Less common nowadays is the use of raptors to hunt rabbits and other game, despite a long-established tradition of falconry in many cultures. The earliest records of falconry as a method of game hunting are from 2000 BC China. Despite its occasional use in controlling crop damage by prey birds, falconry today is largely a hobby.

Cheetahs, too, were used to hunt game by the ancient Egyptians and Assyrians, a practice which became particularly popular during the periods of the Moghul empires of India. Akbar the Great (Moghul emperor AD 1542–1605), kept 9000 cheetahs during his reign and up to 1000 at any one time. Their reluctance or inability to breed readily in captivity is probably the only reason cheetahs haven't been retained as hunting animals.

These examples are not, however, simple cases of exploitation. While humans benefit from the hunting prowess of their companions, the benefits must also be reciprocated. In each case, the animal receives a reward for their services. The fishermen must reward their cormorants with every eighth fish or the birds go on strike and refuse to fish at all. Falcons are given precisely weighted portions of meat after every flight to ensure they remain in peak hunting condition and return to their handler. The cheetahs were always allocated a portion of the kill and, if their hunt was unsuccessful, they returned to their cages voluntarily. But despite such reciprocity there has never been a documented case of a wild dog, cormorant, otter, falcon or cheetah voluntarily entering into a cooperative hunting arrangement with a human. All have been captive animals trained to perform their allotted duties.

A closer parallel with the killer whales and whalers of Eden can be found in parts of Africa where people use a small bird, the aptly named greater honey guide (*Indicator indicator*), to help them find honey. The Boron people of Kenya attract the birds with a special whistle (*Fuulido*) and the birds flit from tree to tree ahead of the humans leading them to a beehive. As they get closer, the bird's behaviour changes, providing their companions with clues as to the proximity of the hive. When the hive is located, the Boron split the hive open, providing access to the honey guides which would otherwise be denied.

Even more similar are cases of cooperation between fishermen and dolphins, which were recorded as early as Roman times. Dolphin–human fishing cooperation is still practiced in areas as far apart as Africa, Australia and South America. Schools of fish close to shore are spotted by the waiting fishermen who 'call' the dolphins by beating the surface of the water rapidly. This sound imitates the noise of the fish leaping out of the water as they are pursued. The dolphins drive the fish shoreward into the waiting nets, spears and arms of the fishermen.

In 1856, Aboriginal people from Stradbroke Island in Queensland were described slapping the water with their spears, eliciting the assistance of local bottlenosed dolphins to drive mullet into their nets. So close was this hunting association, that the Aboriginal people fed the dolphins by hand, recognised individuals and gave them names.

Killer whales, like other dolphins, are also attracted to human fishing activities. Killer whales 'steal' injured whales from Norwegian whalers, and they frequently feed off the carcasses thrown overboard from factory whaling ships in the southern oceans. Their enthusiasm for taking tuna off long-lines in the Indian Ocean has caused fishermen to abandon fishing whenever orcas appeared. In the Bering Sea, pods of up to 50 killer whales appear within hours of hauling operations, taking fish from longlines at depths of 200 or 300 metres. Just recently the Tasmanian blue grenadier fishery has been threatened by local killer whales cleaning off the fishing lines. Exploitation like this provides an easy meal for the killer whales at the fishermen's expense.

Some writers have argued that, far from being beneficial to humans, the Eden killer whales were simply exploiting human hunting. Over-enthusiastic killer-whale activity may even have 'lost' the whalers the odd whale. Waiting to retrieve the carcass when it refloats involved an economic cost to the whalers and increased the risk of losing the carcass in bad weather. In early journal entries, Brierly describes some of the whalers trying to tow the whale ashore immediately and prevent the killers from taking the carcass under water, a strategy also employed later by some of Davidson's competitors. Descriptions of the Eden killer whales as 'dogs', which were rewarded for their efforts by the whalers, further betray a belief that the killer whales were parasiticing the whaler's activities.

Most of the whalers, however, were of the opinion that the killer whales were more help than hinderance. Oswald Brierly believed that the killer whales brought in more whales than they lost. And while some of the short-term whaling stations operating out of Twofold Bay at the height of the whaling boom may have been just as profitable without the killer whales, the longevity of the Davidson whaling station was at least in part due to the assistance of the killer whales.

Historical studies of whaling in Eden have suggested that with low returns from few whales and a chronic shortage of labour, the Davidsons were not able to employ the large numbers of crews required to man multiple boats. With only one or two boats in operation, they would not have been able to retrieve the whales immediately in the face of opposition from the killer whales. It may also have been easier to retrieve a buoyant carcass than a submerged one with a small number of boats.

Furthermore, leaving the whales for the killers to feast upon may have reinforced a relationship which might not otherwise have developed. The Davidson's small-scale whaling operation favoured the killer whales and, in return, the killer-whale activity favoured the Davidsons. Instead of trying to beat the killer whales, the Davidsons joined forces with them.

As well as being a small long-term outfit (rather than the large short-term operation more common in shore-based whaling operations), the Davidsons also operated a very traditional whaling station. Maintaining old-fashioned methods, such as the hand-held lance, had the advantage of reducing overheads in a low-profit environment but also suited the killer whales. The explosive whale guns often used by other whalers, caused the killer whales to abandon the chase, and the Davidsons avoided using them.

> Davidson rarely uses the gun; the killers do not like the noise. He tells a very good yarn on this matter. An opposition whaling station started some time ago, and was getting close to its first whale, used the gun with good result. Shortly after this, Davidson, in yarning with one of the opposition whalers, made a bet that the killers would not help his boat with the next whale. The bet was taken. Next day two whales were sighted. Each station went after a separated whale, but the killers kept with Davidson, who captured his whale. The opponents, not having the assistance of these watchdogs, were unable to get within striking distance of their quarry. It was generally supposed that the killers resented the advent of the opposition whalers, and did all they could to assist Davidson, but as a matter of fact the killers will not work with a boat using guns (Hawkins & Cook, 'Whaling at Eden with some "killer" yarns', p. 270).

The killer whales rapidly learnt to identify the Davidson's characteristic green boats, and while they could not actively choose which whale boat would ultimately kill the whale they had cornered, their loyalty to the Davidson's boats reached legendary proportions.

> On one occasion the opposition boat was beating Davidson in a whale chase. Several of the killers kept swimming back to Davidson, as if to urge him on, while others were lashing the water in front of the opposition boat to impede it. Davidson, knowing the habits and friendly feelings of the killers, as an experiment threw over the painter from the bow of his boat. It was immediately grasped by two killers. They took the rope (tandem fashion) in their mouths, with a half hitch round the shoulder, and started for the quarry. The boat was taken along at a great rate, passing the opposition boat like a flash. About 50 yards from the whale the killers cast off the painter and went back to assist their mates to intercept the other boat. Davidson got fast to the whale and it is the law that whosoever gets fast and remains fast at the death, owns the carcass (Hawkins & Cook, 'Whaling at Eden with some "killer" yarns', p. 270).

While it's unlikely that the killer whales assisted the Davidsons in such an overt manner (and certainly not by grasping the rope in their teeth), they certainly went to great lengths to ensure that the whalers were present at a whale hunt. During the day, the whalers typically kept watch over the bay and surrounding coast from lookouts at Boyd's Tower. The rise behind Kiah Inlet above the whaling station also commanded a fine view of the bay. Perhaps it was for this reason that, while the women of the house used a long-drop toilet nearer to the house, George Davidson preferred the facilities on the top of the rise overlooking the bay. From such lookouts, the whalers soon sighted approaching whales, and particularly those under attack from the killers.

At night, however, whales could not be seen by the whalers. But the killer whales operated under no such restraint and frequently cornered a whale off the bay. Rather than do without the whalers' help, one or two of the killer whales would leave the hunt and head inshore to Kiah Inlet to rouse their assistants. Percy Mumbulla's relating of the oral history of the killer whales, passed on from Aboriginal whalers, echoes that of the white whalers. 'One would go back to Twofold Bay an' leap out of the water. "Pook-urr". He'd slap his tail an' let the whalers know.' (*The Whalers*, p. 16.)

In the mouth of the estuary, the killer whale would floptail to attract the attention of the whalers who, if they hadn't noticed the activity in the bay already, would leap to their boats with a cry of 'Rush-oo' and follow the messenger to the location of the hunt. One contemporary account recorded this activity as follows:

> His 'tally-ho' is sounded by throwing his fifteen to twenty-five feet bulk of bone and blubber clean out of the water, and creating a splash that can be heard for three or four miles. Not having the language of men in his mouth—which is fitted instead with six-inch ivories curved like a rhinocerous horn—he has adopted this signal to let his friends the whale-men know when there is sport afield. (E. J. Brady, 'The law of the tongue', p. 37.)

Despite the scepticism this particular aspect of the killer whale story has been met with, it does appear to be well verified. Eminent marine biologist William Dakin quizzed many former whalers individually about the killer whales 'calling' the whalers to action and was impressed by the consistency and similarity of the descriptions he received from different people, many of whom had long since left Eden. All his informants described the behaviour in the same terms and all referred to it as 'floptailing'. Such consistency convinced the sceptical Dakin that this was a genuine behaviour rather than a myth.

Former Eden Museum curator Alex MacKenzie claimed further that the whalers would slap the water with their oars if they lost sight of the killer whale, and the leader would circle back 'like aquatic sheep dogs, to make sure the dumb humans were still following their lead'. (*The Twofold Bay Story*, p. 55.) Wild killer whales and dolphins are certainly attracted to the sounds of water being slapped, so this story is quite feasible. Other accounts, however, like the claim that the killers placed shells on their blowholes to call the whalers, verge on the absurd.

The habit of the killer whales calling the whalers to the hunt takes their relationship beyond the realms of mutual exploitation. The killer whales not only made use of human whaling activity, they actively solicited human help. Similarly the Davidsons actively encouraged the killer whales by not using explosives, by allowing them to feed on the dead whale and even by feeding them fish in later years when whales were scarce.

Despite the close, long-lasting and relatively well-documented relationship between the Davidsons and the killer whales, the Davidsons were simply the last in a long line of whalers to benefit from the killer whale's activities. Brierly recounts the assistance of the killer whales just fifteen years after the start of the first whaling operation in Twofold Bay. The killer whales then seem to have behaved in much the same way as they did 70 years later towards the end of bay whaling in Eden. The remarkable speed with which this pod developed such a close association with humans bespeaks a far longer, more profound human–killer whale connection, in place well before the arrival of Europeans. It was this earlier relationship that prepared the ground for the cooperative whaling activities, and for this the European whalers had their indigenous crewmen to thank.

*O*nce long ago, all the Ramindjeri people gathered to dance at Moota-paringa (Cut Hill, along Hindmarsh River). They did not have a fire, so they had to dance all day and it was hot. Their perspiration dripped down and became the large ponds there, and hills and valleys formed through the buckling of the ground caused by the stamping of their feet. Eventually they sent for Kondoli, a large and powerful man who possessed fire. He came, but hid his fire. This made the Ramindjeri angry. Another Ancestor, Riballi, threw a spear at Kondoli, hitting him in the neck. The commotion this caused transformed most of the people there into different animals, such as fish and birds. Kondoli himself rushed into the sea and ever after blew water out of his wound. Riballi took Kondoli's fire and placed it in a grasstree (Xanthorrhoea species), where it can be removed by using the dried flower stems as fire-sticks.

The origin of whales in Ramindjeri Aboriginal mythology, as retold in Philip Clarke, 'The significance of whales to the Aboriginal people of southern South Australia', *Records of the South Australian Museum*, p. 23

9

THE KEEPER OF SOULS

The dark people would never go lookin' for whales.
The killers would let them know if there were whales about.
Ole Uncle would speak to them killers in the language.
They must have been bugeens, clever blackfellers...
The killers would only tell the dark people.
The white people had to look for whales themselves.

Percy Mumbulla, *The Whalers*, pp. 16–18

The story of the Eden killer whales and their human helpmates started long before the advent of European whaling. The documented history of the whaling stations of Twofold Bay has, not surprisingly, been a white history, focusing on the owners, operators and managers of the stations—the Imlays, Boyd, Brierly, the Davidsons and others. Far less has been documented about the crews of the stations, the majority of whom were Aboriginal.

Like whaling crews across all oceans, the shore-based whalers of Twofold Bay were a cosmopolitan and itinerant lot—moving with the work and originating from anywhere and everywhere. But workers were in short supply in Twofold Bay, as the Imlay brothers soon discovered in the 1830s. Whaling is hard and dangerous work, not particularly lucrative and provides only seasonal income. The Imlays found a reliable and stable population of workers among the Aboriginal tribes whose land they occupied. They found the local men to be excellent shepherds and cattle handlers, so much so that when one of the Imlay brothers went to New Zealand in 1843, he took two of his best—Aboriginal—shepherds with him. As whalers, their skills and availability made local men a valuable resource. An article by the Sydney surveyor William Henry Wells in the *Gazetteer of the Australian Colonies* (c. 1840) remarked upon the value of south-coast Aboriginal people as a labour force. He commented that the people of Twofold Bay 'are an active and intelligent race, and in their useful labors, in boating and in various arduous employments on board the whalers, certainly contradict the hasty conclusions which so many superficial writers have drawn in reference to the degrading tendency of the faculties of the natives of New South Wales' (quoted by J. A. S. McKenzie, *The Twofold Bay Story*, pp. 26–7). In 1842 Commissioner Lambie reported that three Aboriginal boat crews were employed on the same terms as whites. Aboriginal labour was also used at many other whaling stations around the Australian coast.

Boyd continued the pattern of utilising Aboriginal labour. Oswald Brierly reported that 'all the Twofold Bay blacks are good boatmen, handling their oar with great skill and dexterity, the gins too pull remarkably well' (MLA535: frame G277). He also notes that they were paid in slops and provisions, which some have interpreted as contradicting Lambie's claim that Aboriginal crews were paid the same as white crews. White crews too, however, were typically paid, at least partially, in provisions and food, as was the custom on whaling ships and stations around the world. The value of Aboriginal crews is apparent in the declining number of crews at Boyd's station in 1844. Of the seven crews that started the season, two had left by

September and two followed the next week. Of the three remaining crews, two were Aboriginal. Another station in the same year comprised three Aboriginal crews.

Whatever Aboriginal crews were paid, Boyd did not regard them as the cheapest source of labour available. He attempted to address the chronic labour shortage by 'employing' blackbirded labour from the Pacific Islands. His efforts failed, however, with most of the islanders decamping back home at the first opportunity. Nor were Boyd's efforts appreciated by locals. Humanitarians, especially in Sydney circles, disapproved of such 'slave' labour, while others opposed the scheme for encouraging black, rather than white, immigration. Residents on the south coast found the sudden influx of hungry, cold, fugitive islanders escaping Boyd's workforce so disconcerting that authorities rounded up all the islanders and arranged passage back home.

The gold rushes of the 1850s shrank the pools of available white workers, drawing nearly every able-bodied man off to the goldfields with dreams of fabulous wealth. By the time the Davidson's began whaling, they relied substantially on local Aboriginal people to make up their crew. The Taua language group (also Thaua or Thawa) of Twofold Bay and hinterland was particularly important in providing this workforce. Many of the whalers came from the smaller groups within the Taua tribe, such as the Nullerkermitter from south Twofold Bay and Greencape including Kiah Inlet and the Wiacon from Snug Cove north to Pambula. Access to this reliable workforce was one of the factors that enabled the Davidsons to keep whaling for so long, compared with the short-lived norm for other whaling stations. When the permanent camp of the Aboriginal workers at Kiah Inlet was abandoned later in the early 1900s, whaling virtually ground to a halt. There were simply no crews, skilled or not, to man the boats.

At the height of the whaling industry, Aboriginal crews were not only available, but were successful enough to annoy white crews who sometimes threatened violent reprisals against Aboriginal crews who fastened to a whale before them. Many of the Aboriginal whalers had years of experience and often a family history of whaling, providing finely honed skills which may have been lacking in some of the more itinerant white crews. Keen eyesight and excellent harpoon skills were also great assets for a whaler. But a more significant advantage Aboriginal crews may have had over white crews in the early days was assistance from the killer whales.

Aboriginal oral history suggests that the killer whales favoured Aboriginal crews. Percy Mumbulla, whose Uncle Brierly was a whaler, asserted that the 'killers would only tell the dark people. The white people had to look for whales themselves' (*The Whalers*, p. 18). Such favouritism is not as far-fetched as it might first seem. In the early days of whaling at Twofold Bay, Aboriginal whalers may have been more inclined to leave the whale carcass for the killers to feed on, rather than bringing it straight in to the whaling station. Oswald Brierly's diaries, for example, record an Aboriginal crew telling him that they had marked a humpback which had been taken down by the killers.

In later years, the Davidsons made a regular habit of leaving the whale buoyed and anchored and retrieving it after the killers had taken their fill. This may have been because they lacked the multiple boats required to defend the whale from the killers and tow it back to shore immediately. It may also have been a strategy influenced by the Aboriginal crews who formed such a reliable part of the Davidson workforce. Whatever the origin, this practice is probably the reason why the killer whales were commonly reported to favour the Davidson's green boats over those of their rivals in white boats.

Certainly the killer whales resented any failure to share their spoils and reacted with uncharacteristic violence. Brierly described how the killers would sometimes drag both boat and harpooned whale under water when the whalers attempted to take the whale in immediately. Up to seven boats could be occupied in towing in a whale, while an eighth would have to follow behind attempting to ward off the killers. Even after decades of cooperation, the killers kept a sharp eye on their colleagues. In 1910, a boat attempted to tow a whale back to shore immediately. Tom flung himself up across the back of the boat—much to the consternation of the boat owner who hastily gave up his booty.

Despite not being a particularly maritime people, the Aboriginal people of the New South Wales south coast were excellent harpooners and boatmen. In the early colonial period it was common for Aboriginal men to go to Twofold Bay as harpooners, both for the shore-based stations and on the ocean-going whaling ships that called in at Twofold Bay. Magistrate Alfred Howitt, whose extensive anthropological notes provide important documentation of Aboriginal language and practice in the area, related one local man's story of how his mother had been brought to Twofold Bay to be married and had instead escaped with his father on a whaling ship. Presumably, it was by such means that Aboriginal whalers could be found among the crews of New Zealand ships and stations. Indigenous Australians were yet another nationality among the already highly cosmopolitan crews of the international whaling fleet.

The arrival of Europeans in the early 1800s threw the culture and life of Aboriginal communities along the southeast of Australia into turmoil. With displacement and alienation from land, death by disease and starvation, and cultural corruption by religious and secular authorities, indigenous communities were rapidly damaged. In 1812 several hundred Aboriginal people were reported living in Twofold Bay. By the mid-1800s there were fewer than 100, and at the turn of the century just 10 to 15 remained. Little was recorded of their lifestyles, languages, beliefs and practices by early Europeans and, with the disruption of oral traditions, even the Aboriginal communities themselves lost much of their cultural heritage. That which has been preserved is not always on the public record. Only recently has the importance of Aboriginal crewmen been acknowledged and celebrated as part of Eden's whaling history, particulary by the research and documentation of Sue Wesson.

The paucity of information about south-eastern Aboriginal associations with whales before and during early European contact can partly be supplemented by information from other coastal communities. Tribes around Sydney left tangible evidence of their interest in whales engraved in the rocks on headlands throughout the area. Early settlers in other parts of the country, such as Queensland and South Australia, sometimes recorded more extensive details of Aboriginal associations with whales than exist for the Twofold Bay area. Recent research by Philip Clarke of the South Australian Museum, for example, has brought together a wealth of information on whaling and whale associations of the Aboriginal people of the Lower Murray region. Their history might shed some light on the kinds of beliefs and practices that might have existed in Eden.

There is no evidence to suggest that Aboriginal people actually hunted whales before European settlement, but there are many accounts of whales being eaten. Whale strandings along the coast were probably common enough to be regarded as welcome bounty and cause for a celebratory corroboree among neighbouring tribes. A rock engraving at Cowan near Sydney depicts such a feast around a stranded whale.

A number of early European explorers commented upon the consumption of whales. When Matthew Flinders surveyed the coast with George Bass, whale was the source of an interesting observation on cross-cultural tastes and etiquette.

> October 9th, 1798. On the way from Snug Cove through a wood to the long northern beach where we proposed to measure a base-line, our attention was suddenly called by the screams of three women, who took up their children and ran off in great consternation. Soon afterwards a man made his appearance. He was of middle-age, unarmed except for a waddie or wooden scimitar, and came up to us seemingly with careless indifference. We made much of him and gave him some biscuit, and in return, he presented us with a piece of gristly fat, probably of a whale. This I tasted, but watching an opportunity to spit it out when he should not be looking, I perceived him doing precisely the same thing with our biscuit, whose taste was probably no more agreeable to him than his whale was to me (quoted in J.A.S. McKenzie, *The Twofold Bay Story*, pp. 6–7).

Whale meat, blubber and oil from a stranded animal would have provided plenty of food for many people, particularly in the otherwise lean winter months when food was more difficult to acquire along the coast. In South Australia, cooked whale meat was eaten with pigface leaves (a succulent coastal plant).

Whale earbones were used to store water and the giant ribs were sometimes used as the framework for shelters. This use of whalebones in home construction was later imitated by European whalers. The Imlay's houses in Eden were characterised by the use of the large whale vertebrae as paving for garden paths.

In southern New South Wales, whale oil was used for medicinal and ceremonial purposes. Oil was extracted from the whale by placing pieces of blubber on the edge of a hollowed-out rock pool above the water line. A fire was then lit on the high side of the rock pool, heating the blubber and enabling the oil to be squeezed out and collected in the pool below. Such extraction sites were sacred to the local people and of great ceremonial significance. Aboriginal people from the Lower Murray in South Australia rubbed whale oil on themselves as protection fromthe cold in winter and combined it with red ochre to decorate themselves.

Whale carcasses were also valued for their therapeutic effect as a cure for rheumatism and arthritis. The combination of heat generated by the rotting carcass and oil apparently offered some relief to those brave enough to encase themselves within the carcass. This practice may even have been illustrated in rock-carvings at Ball's Head, Allambie Heights and West Head in Sydney. These images of human figures drawn inside a whale outline could illustrate Dreaming stories of whales consuming humans, humans riding whales, or other interactions between human and whale spirit creatures. The figures could, however, also illustrate the medicinal use of whale carcasses. In the late 1800s, this traditional Aboriginal treatment spread to the white community. The whaling stations of Eden became one of the more unusual destinations of those desperate souls seeking a cure for painful joint conditions. In 1908, Hawkins and Cook, described the following unpleasant treatment:

> A number of persons suffering from rheumatism visited Eden in the whaling season for a 'whale cure'. The body of the whale retains a certain heat long after death—a sort of fermentation heat. A hole is dug in the blubber with spades into the whale's inside, deep enough to hold the body of a man or woman: into this hole the sufferer is lowered, and remains in as long as he can stand the heat—and smell. Some remarkable cures have been effected which in the past had baffled all medicinal and other treatments. The after-effects are not so pleasant; the patient for a week or so gives off a horrible odor, and is abhorrent to man and beast, and a fit subject for prosecution under the 'Diseased Animals and Meat Act' ('Whaling at Eden with some "killer" yarns', pp. 271–2).

The famous rheumatism cure. George Davidson supervises the treatment.

Whales were highly significant animals in many Australian coastal communities. The southern right whales were regarded as totemic ancestors by many people living along the Lower Murray coastal area. In their accounts, the whale ancestor, Kondoli, typically has the ability to make fire but will not share it. After a fight, Kondoli escapes to the sea, sometimes with a spouting wound, and his fire-making ability is transferred to the grass-tree, which is commonly used to start fires, or to flints in the ground. In other accounts, the ability to make fire is stolen by the shark man, whose flints then turn to teeth.

The relationship between whales and fire probably stems from the smoke-like eruptions from the whale's blow-hole and the fact that whales, unlike fish, are warm-blooded. Rotting whale carcasses also generate a considerable amount of heat. The similarity between whale blows and smoke was also noted by European whalers, who described a wounded whale's bloody blows as having 'its chimney on fire'—a condition indicative of impending death.

The benefits of stranded whales gave rise to the belief that they had been driven ashore deliberately. Some coastal tribes believed the whales were stranded by friendly bird-like spirits. On the southeast coast of South Australia, indigenous people believed that whales could be sung ashore. The coast of Younghusband Peninsula, south of the mouth of the Murray, is renowned for whale strandings. Various whale songs have been recorded from this area. One chant in the Booandik language (transcribed in 1880) translates as 'The whale is come. And thrown up on land', a refrain repeated at length to induce a stranding. Such whale-singing skills were generally confined to members of clans descended from the whale ancestor Kondoli, and these people largely controlled the distribution of whale meat, although they did not always eat their totemic ancestor themselves. Some people believed that the whales were brothers returning to their homeland to die. Once their spirit had returned home, the meat could be eaten by their family. Other accounts suggest that whale-descent groups tried to prevent the whales from stranding by singing them *off* the shore. The remains of a totemic ancestor were potent magic against its descendants, so while whale-descent clans were careful to ensure that all significant whale remains were used or hidden, other groups used the whale oil as 'spear-poison' against people with the whale as their totem.

At Eden, however, it was the killer whales that drove the whales ashore. Aboriginal people probably benefited from the killer whales' activities when whales stranded in an effort to escape their relentless pursuers, or when the whales washed ashore after the killers had partaken of their favoured morsels—the lips and tongue. But the relationship between local Aboriginal people and killer whales may have been closer than such occasional exploitation of the killer whale's hunting prowess.

Only a handful of early anthropologists attempted to record the details of traditional pre-European culture in this area. One such record comes from retired surveyor, Robert Hamilton Mathews, who spent the last twenty years of his life documenting the traditional life and customs of many Aboriginal people. Mathews collected the following story from the Twofold Bay area in the early 1900s:

> When the natives observe a whale 'mūrirra,' near the coast, pursued by 'killers,' mánanna, one of the old men goes and lights fires at some little distance apart along the shore, to attract the attention of the 'killers.' He then walks along from one fire to another, pretending to be lame and helpless, leaning upon a stick in each hand. This is supposed to excite the compassion of the 'killers' and induce them to chase the whale towards that part of the shore in order to give the poor old man some food. He occasionally calls out in a loud voice, ga-ai! ga-ai! ga-ai! Dyundya waggarangga yerrimaran-hurdyen, meaning 'Heigh-ho! That fish upon the shore throw ye to me!
>
> If the whale becomes helpless from the attack of the 'killers' and is washed up on the shore by the waves, some other men, who have been hidden behind scrub or rocks, make their appearance and run down and attack the animal with their weapons ('Ethnological notes', pp. 252–3).

The killer whales' habit of driving small whales into the shallows benefited the local Aboriginal people greatly and they may have encouraged this behaviour by calling to the killer whales, smacking the water and offering them fish.

Killer whales were certainly important animals in the spiritual lives of the Taua people. As we have learnt, these *beowas* or *mánanna* were regarded as the reincarnated spirits of past tribesmen, particularly warriors or, in later years, whalers. It has even been suggested that the black-and-white corroboree dress traditional to the area is related to the striking markings of the killer whales.

Brierly repeatedly documented the Aboriginal belief that the killer whales took on the souls of Aboriginal whalers who had died:

> ...the natives of Twofold Bay regard these killers as the reincarnate spirits of their own departed ancestors—and so firm is their belief that they go so far as to particularise and identify certain individual killer spirits. Hence I had the greatest difficulty in obtaining the head of one of these animals as...the men had the greatest reluctance at risking all luck by killing a specimen (*Reminiscences of the Sea: About whales* MLA546: frame 10).

The belief that the killer whales harboured the souls of warriors who had died extended beyond just the Aboriginal community. An article in the *Argus* in 1948 described how 'many local residents in the early days claimed that it was a remarkable coincidence that when one of the local Aboriginal people died a new killer appeared in the bay shortly after'.

While the killer whales were generally regarded by the indigenous community as friendly creatures, which did not harm humans, they were also held in awe by the Aboriginal whalers. Brierly recounts how the Aboriginal crews refused to put out in the boats after one of their colleagues was knocked out of the boat and killed. The day before a killer whale had been sighted whose soul was that of a hostile neighbouring tribesman. Latter-day accounts of the Eden killer whales (even by Aboriginal authors) tend to accord them the status of working dogs— friendly, playful and cooperative. Percy Mumbulla, for example, related how the Aboriginal whalers 'chuck a big lump of blubber to the killer. He's like their dog.' (*The Whalers*, p. 14). Tom Mead reported that George Davidson 'had come to regard these killers, and Tom in particular, more or less as pet dogs' (*Killers of Eden*, p. 145). Over time perhaps the killer whales' image has been somewhat sanitised. I suspect that the actual role of the killer whales in Aboriginal culture was far more ambivalent. The killer whales were keepers of souls—a portal to the 'otherworld' of death. They may have been favourably disposed towards humans, but like any other spiritual creature, they demanded respect.

The spiritual relationship between the Taua and the killer whales may have been similar to relationships between killer whales and indigenous cultures in other parts of the world. A strong association with the natural land and its creatures (such as in hunter-gatherer cultures) lends itself to animistic belief systems in which soul exchange is a pivotal feature. In such religions, there is a constant flow of souls back and forth between humans and animals, plants and geographic features. Great significance is attached to the journey of these souls between their resting places and the spirit world and ensuring that they are not lost in transit. The sea, the sky and the night are often seen as containing spirit worlds. Creatures capable of crossing from the human day-lit terrestrial world into these other—spirit—dimensions are essential for soul journeys. Birds, for example, cross the aerial/terrestrial boundary. In some Aboriginal cultures, black cockatoos accompany souls on their journey home. Nocturnal creatures, like owls and bats, cross the boundary between night and day—the European mythology associated with vampires and bats clearly stems from a preoccupation with soul transfer and soul stealing into the night. In coastal cultures, sea mammals, like whales and seals, are important messengers from the aquatic underworld. Dolphins are messengers between this world and the next in Amerindian culture. Interestingly, while terrestrial soul keepers are typically masculine, in the aquatic world they are often female figures. The famous sirens who lured sailors to their deaths and the 'selkie' seal-women of Celtic mythology are both concerned with the capture and transfer of souls to and from the undersea world.

The status of chief soul keeper is often accorded to a top predator. The daily dispensing of death (and by default, life) make predators natural candidates for playing a judge-like role in the distribution of souls. In cultures across the world, the jaguar, the polar bear, the wolf and the tiger are regarded as soul keepers, and where cultures are strongly associated with the sea, a marine predator like the killer whale becomes soul keeper. The Latin name for killer whale *orcinus* (meaning 'of or belonging to the realms of the dead') is sometimes related directly to the killer whale's predatory ability, but it is more likely to originate in this deep-rooted spiritual belief system of major predators as the keepers of souls.

For the Nazca people of Peru, killer whales were symbols of power, warrior courage and fertility. Nazca temples constructed around 100 BC, were dedicated to killer-whale deities and their icons were decorated with stylised killer-whale motifs. To the Haida Gwaii of northwest Canada, the killer whales are *skana* or demons, with supernatural powers. The Haida have a particularly rich killer-whale culture, much of which is preserved in artefacts and totemic images. Totem poles, bowls, blankets and houses are all richly decorated with the images of Haida spiritual life, in which killer whales feature strongly. The Haida live on Queen Charlotte Island, in the Johnstone Strait off British Columbia. The Johnstone Strait is famous for its salmon runs, with numerous species of salmon migrating up the rivers on both sides of the Strait to breed. The seasonal flow of the salmon up and down the rivers plays a pivotal role in the lives of both the Haida and the resident pods of killer whales in the Strait.

The Haida divide into two social groups, or moieties, the Raven and the Eagle, each of which split into more than twenty lineages. Each lineage has a crest featuring a mythical being or animal; killer whales being particularly popular among the Raven lineages. The 'raven-finned killer whale' crest, for example, refers to a story in which the legendary Raven pecked himself out of a killer-whale body through the end of its dorsal fin. Killer whales also feature in Eagle lineages, as expressed in the 'five-finned killer whale' crest. These crests link the lineages to specific killer-whale chiefs from nearby undersea villages. These villages are direct counterparts of the villages on land, reinforcing a reciprocal relationship between killer whales and humans.

As transitional creatures breathing air, yet living in water, killer whales were seen by the Haida as being of two worlds. Their prominent dorsal fins, breaking the surface of the water, are a doorway to the underworld. Images of killer whales commonly depict human figures—demons and shamans—peering out of their fins, and often feature on 'soul-catcher' tools, used to capture lost souls that may have wandered from a sick person. Restoring an errant soul returns a person to health, and soul-catchers were often built into the chimneys and doorways of houses while portable soul-catchers were used by the local shaman.

The mix of fear and respect killer whales commanded was also transmitted to the loggers and other settlers who arrived in British Columbia. One logger related how he and his colleagues were pushing logs down into the water when a pod of killer whales passed close by the shore. His friend pushed a log down, deliberately striking an orca and injuring it. That night as the loggers rowed back, their boat was capsized by the whales. The man who had pushed the log disappeared, while the narrator lived to tell the tale.

The significance of killer whales continues in communities further west into the Pacific. On the Aleutian Island of Alaska, killer whales are known as *polossatik*, or 'the feared ones'. The Tlingit of Alert Bay on the Alaskan panhandle actively avoid them, believing that the killers will wreak vengeance on anyone who kills a member of their pod. An Inuit story tells of a man who harpooned a killer whale and each time he went to the edge of the ice to launch his kayak he found the killer whales waiting for him. In the end he had to give up the sea. Similar spiritual significance is accorded to killer whales in the traditional coastal cultures of Russia, while the Ainu of Japan call the whale *repun kamui*, or 'master of the open sea'.

In 1953, Alaskan native hunters were seen harvesting beluga whales that were being driven ashore by killer whales, just as the Twofold Bay Taua people seem to have done centuries beforehand on the opposite side of the world. The potential for cooperation between humans and killers may well have existed at other times and places, but only in Eden did it come to its full fruition. European whaling brought the relationship to a professional peak and demonstrated an appetite for baleen whales matched only by that of the killer whales themselves. But the foundation for the partnership lies firmly in the centuries-old spiritual and emotional bond between the local Aboriginal people and the killer whales. Only George Davidson's friendship with Tom seems to have come close to recalling the depth of the bond between the Aboriginal whalers and the Eden killer whales. It was this bond which was responsible for the unique development of cooperative hunting between killer whales and humans in Twofold Bay. These whales were not just helpers, allies or workmates. They were not merely friends. The killer whales were family—brothers and ancestors—the living embodiment of souls of departed relatives. Just as kinship is central to the maintenance of killer whale society, the Aboriginal whalers cemented their bond with the killer whales in the blood ties of their own kinship system.

It is a curious coincidence that when the Nullica tribe departed from Kiah Inlet in the early 1900s, the bulk of the killer whales also left Twofold Bay for the last time.

Two mighty whales! Which swelling seas had tost,
And left them prisoners on the rocky coast;
One as a mountain vast, and with her came
A cub, not much inferior to his dam.
Here in a pool, among the rocks engaged,
They roared, like lions caught in toils, and raged

Quoted in D.G. Stead, *Giants and Pigmies of the Deep*, p. 21
in reference to the stranding of a 98-foot (30 m) long blue whale
and its 45-foot (14 m) long offspring at Twofold Bay.

EPILOGUE

Soon as ever the dark people left Twofold Bay an' come to Wallaga Lake, them killers went north, because there were no blackfellers there.

Percy Mumbulla, *The Whalers*, p. 30

The disappearance of the killer whales from Eden, the fame surrounding Tom's death and his immortalisation in the Eden Killer Whale Museum give the impression that the Twofold Bay killer whales somehow 'died out'. Some locals believed that off-shore whaling fleets had killed many of the killer whales during the summer months. Certainly, that was the fate of the great baleen whales of the area, so perhaps the predators who depended upon them for food suffered a similar fate?

As top predators, killer whales might seem somehow invulnerable to the mortality factors affecting lesser creatures further down the food chain. But like top predators everywhere, killer whales are, in fact, highly susceptible to a wide variety of environmental changes. The top of the food pyramid is a very precarious position and no amount of teeth and muscle can ward off some of the dangers. Any disruption in the food chain further down the line can dramatically affect the predators who depend on other animals for food.

The wholesale slaughter of the baleen whales must have had a huge impact on the killer whales that specialised in eating them. In the early days of whaling, Oswald Brierly observed blue whales in a huge school stretching as far as the eye could see. By the end of Brierly's life, such super-pods were unheard of. Some species, like the blue whale, have been reduced to a mere handful of animals scattered widely across empty oceans. The ocean-going whale-killers must have noticed the dramatic decline in food even over the course of their own lifetimes.

Killer whales specialising in other food, of course, would have been unaffected by this particular act of human excess, so killer whales worldwide remained abundant. And it is possible that the affected whale-killers may have diversified their diets to take in other food sources, such as smaller whales like the minke, seals and other marine mammals. The difficulty and danger of capturing large baleen whales probably means that they are only ever an occasional treat for killer whales and must be supplemented by smaller, more reliable prey items. Nonetheless, killer whales do not seem to like changing diets, so the decline in baleen whales may well have made whale-killing families less successful. In the Crozet Archipelago, a decrease in the elephant seal population over the last 25 years appears to have resulted in a steady 4 per cent annual decline in the numbers of killer whales that eat them. With less food, whale-hunting killer whales may also have found it more difficult to raise young and, over time, families like the Eden killer whales might well have died out.

Human predation also directly affects killer-whale populations. Fish-eating whales are still targeted by fishermen in oceans around the world. Up to 25 per cent of live-caught killer whales in the Puget Sound, Washington State, have pre-existing bullet wounds. Killer whales in more isolated areas (such as open oceans and Antarctica) are generally less likely to face high levels of attack on an ongoing basis, but, even in the open ocean, they can come under periods of sustained hunting pressure—as witnessed in Iceland in the 1950s, Norway in the 1970s and more recently in Antarctic waters. The Eden killer whales were renowned for refusing to assist whalers who used an explosive whale gun. They may well have experienced such devices turned against themselves in the past and were not willing to risk working near boats so equipped. Killer whales off Japan in the 1970s rapidly learnt to avoid those fishing boats with mounted guns and concentrated their efforts to 'steal' fish around unarmed fishing boats, despite the fact that the boats all looked superficially similar. At least two of the Eden killer whales, Typee and Stranger, appear to have been killed deliberately by humans. Others may also have perished at human hands particularly once they left the protected environment of Twofold Bay. After their departure, rumours abounded in Eden that the killer whales had been shot while following larger offshore whaling ships because of the chaos they caused during a whale hunt.

While their ocean-going counterparts are less vulnerable to human activities, coastal killer whales are particularly susceptible. Whales are occasionally struck by boats, indeed, 1 per cent of bowhead whales harvested from the Arctic seas around Alaska show signs of having survived being struck by a ship. Similarly, killer whales in New Zealand and British Columbia have been identified bearing the scars of propellers. It seems likely that mature whales accustomed to power boats soon learn to avoid them, and that injuries are more likely to be sustained by inexperienced calves and adults from areas without regular boating activity. The captain of a British Columbian ferry described the following incident in the early 1970s:

At 3.45 there was a crunch at the after end of the ship, as if we had struck a small log. I went and looked out the winder at the back of the wheelhouse and noticed a reddish-brown discolouration in our wake. My first impression was that we had struck a butt-end of a dead-head just below the surface of the water. Then four killer whales surfaced about two to three ship lengths astern.

The first thing I noticed about these four surfacing whales was that one was bleeding profusely. I told the Quartermaster to bring the ship hard around and we steamed up to within ten feet of the whales. The pod consisted of a bull, cow and two calves. It was one of the calves that had been struck by the ship's propellers. It was a very sad scene to see. The cow and bull cradled the injured calf between them to prevent it from turning upside-down. Occasionally the bull would lose its position and the calf would roll over on its side. When this occurred the slashes caused by our propeller were quite visible. The bull, when this happened, would make a tight circle, submerge and rise slowly beside the calf, righting it, and then proceed with the diving and surfacing. While this was going on the other calf stayed right behind the injured one.

We stayed with the whales for about ten to fifteen minutes; there was no fear of the ship being too close (about ten feet at times). I felt at the time that there was very little we could do to alleviate the obvious pain and suffering that was taking place and that the calf could not survive too long (cited in Ford, Ellis & Balcomb, *Killer Whales*, p. 83).

Such direct killings, however, are probably less significant than the insidious dangers of living near humans. Recently, scientists have found disturbingly high levels of pollutants in blubber samples from British Columbian killer whales. Concentrations were particularly high in older males and in the transient killer whales which specialise in eating marine mammals. Since toxins concentrate as they move up the food chain, increasing in each predator, seal-eating killer whales are particularly vulnerable to contamination. For some reason, reproductive activity appears to protect the females from the worst of the contamination.

Such modern threats to killer-whale survival are, however, unlikely to have been responsible for the disappearance of the Eden killer whales. It is more likely that these killer whales simply started spending their winters in more lucrative hunting grounds as the baleen whales became scarce.

Reduced food may not have been the only reason the killer whales left Eden. After Typee was killed in 1901 while stranded on Aslings' Beach, the number of killer whales returning to Twofold Bay substantially declined. It seems quite likely that the sub-pod comprising Typee's immediate family (probably his mother and siblings) never returned to Twofold Bay after his death, leaving just the two remaining sub-pods in residence.

By 1912 the number of killers present in Twofold Bay had declined to seven, suggesting that a second sub-pod, possibly Stranger's family, had failed to return. Stranger's death in 1907 at the hands of a fisherman may well have greatly disrupted her family's behaviour and migrations.

The remaining family group of seven whales was presumably Hooky's family which included Tom and perhaps Kinscher, Charlie Adgery and Young Ben. By 1923 just three whales returned each year to Twofold Bay, including Tom, Hooky and another animal, perhaps Humpy and probably Tom's sibling. This remaining small family was probably comprised of elderly animals—certainly Hooky and Tom were quite old by this stage, judging from the period of time over which they had been iden-tified in Eden. Perhaps these last few animals were too old to take up new hunting habits and continued to return to the familiar confines of Twofold Bay in their later years.

Despite the deaths of Stranger, Typee, Tom and others, it seems unlikely that the pod as a whole perished. More probably, it continued hunting the great baleen whales up and down the east coast of Australia. In the 1930s, for example, Sydney residents witnessed a series of spectacular attacks on humpback whales by killer whales:

> During each whale season on the New South Wales coast, these attacks on the whales by the Killers are witnessed from ships and frequently from the land itself. The 1930 season (from June till October) was particularly marked by a number of combats. Some of these were actually witnessed by hundreds of residents of the ocean-side suburbs of Sydney. But—most unique of all—we had within the space of one month, two recorded instances of air travellers by the Sydney–Brisbane air mail being able to fly over and observe at fairly close quarters the struggles of a mighty whale against its dreadful tormentors (D.G. Stead, *Giants and Pigmies of the Deep*, p. 39).

Given the scarcity of killer whales along the Australian coastline, and whale-hunting specialists in particular, I think these unusual attacks were probably the Eden killer whales in operation. With the lower density of baleen whales moving up and down the coast, the killers may have started actively seeking out whales, rather than simply ambushing passing whales at Eden.

Long-distance travel is well within normal killer whale behaviour—particularly for ocean-going or whale-hunting populations. While many killer whales, such as the resident salmon-eating pods of British Columbia, operate within relatively small ranges of around 40 to 60 square kilometres, ocean-going killer whales probably travel quite long distances. A number of killer whales attacking a gray whale mother and calf in Monterey Bay, California, had been photographed three years earlier in Glacier Bay, Alaska—some 3000 kilometres north along the coast. Antarctic killer whales have likewise been sighted in the Bay of Islands in northern New Zealand.

The killer whales seen up and down the southeastern Australian coast each winter today are probably members of just a handful of families, which regularly patrol the area. They are most commonly seen from June to November off Eden and are sighted each September off Port Stephens, north of Sydney. Further north the baleen whales veer offshore outside the Great Barrier Reef on their way to breeding grounds in the Coral Sea, probably taking their predatory followers with them.

These killer whales may well be descendants of the Eden killer whales. But, just as there are no whalers alive today who recall first-hand the killer whale–human collaboration, there may well not be any individual killer whales among these families who recall those times either. Among the whaling community, it tends to be women (with their longer lifespans) who still remember the killer whales. George Davidson's daughters, who observed the activities of the whales from shore and recall many of the family stories associated with the killer whales, are in their nineties now. Margaret Brooks, also in her nineties, accompanied her father John Logan, on his launch in the last few years of whaling and observed at close quarters the last of the killer-whale hunts.

Like humans, female killer whales may also be the repositories of collective memories. If the pods of killer whales currently traversing the east coast of Australia really are descended from the Eden killer whales, perhaps there is an ancient clan grandmother who vaguely recalls the glory days of her parents hunting huge numbers of baleen whales off Twofold Bay. Perhaps it is not entirely by chance that killer whales are occasionally seen in their old haunt off Leatherjacket and Twofold Bay. Maybe they return with the vague recollection that once upon a time this was a good hunting ground. But, then again, if killer whales are capable of such complex thoughts, the memories of the days when humans helped them catch their prey may well and truly have been relegated to the stuff of fantastic and improbable orca legend.

I never saw a killer whale in Eden. Not on my first visit as a child. Not when I returned twenty years later to research this book. In fact, in all the years spent sailing along the Australian coast, studying seabirds on remote islands in the Scottish Hebrides, visiting Iceland or holidaying on the coast, killer whales have eluded me. My friends have seen them. They were in the harbour yesterday, two weeks ago or around the corner in the next bay. I've read about killer whales, studied them, watched films about them, written scientific papers on them, lectured on them, even dreamed about them but I've yet to actually see a killer whale in the flesh. For now, the killer whales I know best are the ones I can only ever meet in my imagination—Tom, Hooky, Humpy, Cooper, Typee, Jackson, Stranger, Big Ben, Young Ben, Jimmy, Kinscher, Sharkey, Charlie Adgery, Brierly, Albert, Youngster, Walker, Big Jack, Little Jack, Skinner and Montague. This has been their story as best I can reconstruct from a distance of decades. I hope I've done justice to their remarkable lives.

FURTHER READING

General references

Brady, E. J. (1909) 'The law of the tongue: Whaling, by compact, at Twofold Bay', *Australia Today*, 1 December: 37–9.

Brierly, O. (1842–8) *Diaries at Twofold Bay and Sydney*, State Library of New South Wales, MLA503–541.

Brierly, O. (1844–51) *Reminiscences of the Sea: About whales*, State Library of New South Wales (Mitchell Library Archives), MLA546.

Brierly, O. (1842–3) *Journal of a visit to Twofold Bay, Dec. 1842–Jan. 1843*, State Library of New South Wales, MLA535.

Dakin, W. J. (1938) *Whalemen Adventurers*, Angus & Robertson: Sydney.

Davidson, R. (1988) *Whalemen of Twofold Bay*, René Davidson: Eden.

Hawkins, H. S. & Cook, R. H. (1908) 'Whaling at Eden with some "killer" yarns', *Lone Hand*, 1 July, 3: 265–73.

McKenzie, J. A. S. (undated) *The Twofold Bay Story*, Eden Killer Whale Museum and Historical Society: Eden.

Mead, T. (1961) *Killers of Eden*, Angus & Robertson: Sydney.

Mitchell, M. (undated) *Whale Killers of Twofold Bay*, Eden Killer Whale Museum: Eden.

Mumbulla, P., Robinson, R. & Bancroft, B. (1996) *The Whalers*, Angus & Robertson: Sydney.

Stead, D. G. (1933) *Giants and Pigmies of the Deep: A story of Australian sea denizens*, The Shakespeare Head Press: Sydney.

Wellings, C. E. (1944) 'The killer whales of Twofold Bay, NSW Australia, *Grampus orca*', *Australian Zoologist*, 10: 291–4.

Wellings, H. P. (1964) *Shore Whaling at Twofold Bay: Assisted by the renowned killer whales*, Eden Killer Whale Museum: Eden.

Wesson, S. (2000) *A history of Aboriginal involvement in whaling at Twofold Bay*, report for the New South Wales National Parks & Wildlife Service and Bega, Eden & Merrimans Aboriginal Forests Management Committee.

Other references

Chapter Two—Demon Dolphins

Baker, A. N. (1990) *Whales and Dolphins of Australia and New Zealand: An identification guide*, Allen & Unwin: Sydney.

Corkeron, P. J. & Connor, R. C. (1999) 'Why do baleen whales migrate?' *Marine Mammal Science*, 15: 1228–45.

Evans, P. G. H. (1984) 'Whales and Dolphins', in Macdonald, D. W. (ed.) *The Encyclopaedia of Mammals*, Unwin Hyman: London.

Florezgonzalez, L., Capella, J. J. & Rosenbaum, H. C. (1994) Attack of killer whales (*Orcinus orca*) on humpback whales (*Megaptera novaeangliae*) on a South American Pacific breeding ground, *Marine Mammal Science*, 10: 218–22.

Forney, K. A. & Barlow, J. (1998) 'Seasonal patterns in the abundance and distribution of California cetaceans', 1991–1992, *Marine Mammal Science*, 14: 460–89.

Gould, S. J. (1993) *The Book of Life*, Random House: Australia.

Macdonald, D. W. & Barrett, P. (1993) *Mammals of Britain and Europe*, HarperCollins: London.

Martin, A. R. (ed.) (1990) *Whales and Dolphins*, Bedford Editions: London.

Nichol, L. M. & Shackleton, D. M. (1996) 'Seasonal movements and foraging behaviour of northern resident killer whales (*Orcinus orca*) in relation to the inshore distribution of salmon (*Oncorhynchus* spp.) in British Columbia', *Canadian Journal of Zoology*, 74: 983–91.

Simila, T., Holst, J. C. & Christensen, I. (1996) 'Occurrence and diet of killer whales in northern Norway: Seasonal patterns relative to the distribution and abundance of Norwegian spring-spawning herring', *Canadian Journal of Fisheries & Aquatic Sciences*, 53: 769–79.

Thewissen, J. G. M. (1998) *The Emergence of Whales; Evolutionary patterns in the origin of Cetacea*, Plenum Press: New York.

Ursing, B. M. (1998) 'Analyses of mitochondrial genomes strongly support a hippopotamus-whale clade', *Proceedings of the Royal Society of London, Series B*, 265: 2251–5.

Vickers–Rich, P. & Rich, T. (1993) *The Wildlife of Gondwana*, Reed: Sydney.

Visser, I. N. (1999) 'Antarctic orca in New Zealand waters?', *New Zealand Journal of Marine & Freshwater Research*, 33, 515–20.

Chapter Three—The Baleen and Blubber Boom

Anon. (1872) 'Whaling in Twofold Bay', *Illustrated Sydney News and NSW Agriculturalist & Grazier*, 26 October, p. 3.

Anon. (1903) 'Whaling at Eden', *Sydney Mail*, 7 January.

Davidson, D. (undated) 'The Davidsons of Kiah Inlet', unpublished family history, Eden Killer Whale Museum archives.

Ford, J. K. B., Ellis, G. M. & Balcomb, K. C. (1994) *Killer whales: The natural history and genealogy of* Orcinus orca *in*

British Columbia and Washington State, University of British Columbia Press, Vancouver.

Mabey, R. (ed.) (1995) *The Oxford Book of Nature Writing*, Oxford University Press, Oxford.

Waitt, G. & Hartig, K. (1997) 'Grandiose plans, but insignificant outcomes: the development of colonial ports at Twofold Bay, New South Wales', *Australian Geographer*, 28: 201–18.

Chapter Four—Old Tom

Graham, M. S. & Dow, P. R. (1990) 'Dental care for a captive killer whale, *Orcinus orca*', *Zoo Biology*, 9: 325–30.

Mitchell, E. & Baker, A. N. (1980) 'Age of reputedly old killer whale. *Orcinus orca*, "Old Tom" from Eden, Twofold Bay, Australia', *Report of the International Whaling Commission* (special issue *3*): 143–54.

Tomes, C. S. (1914) *A Manual of Dental Anatomy: Human and comparative,* (7th ed), J. & A. Churchill: London.

Chapter Five—A Matter of Taste

Aelianus, C. (1958) *On the characteristics of animals*, Vol. XII, Harvard University Press: Cambridge.

Andriano, J. (1999) *Immoral monster: The mythological evolution of the fantastic beast in modern film and fiction*, Greenwood Press: Westport, Conneticut.

Bagshawe, T. W. (1939) *Two Men in the Antarctic: An expedition to Graham Land 1920–22*, Cambridge University Press: Cambridge.

Cherry-Garrard, A. (1922) *The Worst Journey in the World, Antarctia 1910–1913*, vol. 1, Constable & Co, London.

Clarke, J. (1969) *Man is the Prey*, Stein & Day: New York.

Ellis, R. (1994) *Monsters of the Sea*, Knopf: New York.

Hoyt, E. (1990) *Orca: The Whale Called Killer*, Camden House: New York.

Jefferson, T. A., Stacey, P. J. & Baird, R. W. (1991) 'A review of killer whale interactions with other marine mammals—Predation to coexistence', *Mammal Review*, 21: 151–80.

Kasamatsu, F. & Joyce, G. G. (1995) 'Current status of odontocetes in the Antarctic', *Antarctic Science*, 7: 365–79.

Kipling, R. (1902) 'How the whale got his throat', *Just So Stories*, Macmillan: London.

Scott, R. F. (1913) *Scott's Last Expedition*, vol. 1, Smith, Elder & Co.: London.

Snorf, C. R., Hughes, J. & Hattori, T. (1975) 'Killer whale attack on a surfer: A case report', *Journal of Bone & Joint Surgery*, 57–A: 138.

Chapter Six—An Eelectic Palate

Baird, R. W. & Dill, L. M. (1995) 'Occurrence and behavior of transient killer whales—Seasonal and pod-specific variability, foraging behavior, and prey handling', *Canadian Journal of Zoology*, 73: 1300–11.

Baird, R. W. & Stacey, P. J. (1989) 'Observations on the reactions of sea lions, *Zalophus californianus* and *Eumetopias jubatus*, to killer whales, *Orcinus orca*—Evidence of prey having a search image for predators', *Canadian Field Naturalist*, 103: 426–8.

Constantine, R., Visser, I., Buurman, D., Buurman, R. & McFadden, B. (1998) Killer whale (*Orcinus orca*) predation on dusky dolphins (*Lagenorhynchus obscurus*) in Kaikoura, New Zealand, *Marine Mammal Science*, 14: 324–30.

Domenici, P., Batty, R. S., Simila, T. & Ogam, E. (2000) 'Killer whales (*Orcinus orca*) feeding on schooling herring (*Clupea harengus*) using underwater tail-slaps: Kinematic analyses of field observations', *Journal of Experimental Biology*, 203: 283–94.

Eschricht, D. F. (1866) 'On the species of the genus Orca

inhabiting the northern seas', in *Recent Memoirs of the Cetacea* (W. H. Flower, ed.) London: Ray Society, pp 151–88.

Fertl, D. & Acevedo-Gutierrez, A. (1996) 'A report of killer whales (*Orcinus orca*) feeding on a carcharhinid shark in Costa Rica', *Marine Mammal Science*, 12: 606–11.

George, J. C., Philo, L. M., Hazard, K., Withrow, D., Carroll, G. M. & Suydam, R. (1994) 'Frequency of killer whale (*Orcinus orca*) attacks and ship collisions based on scarring on bowhead whales (*Balaena mysticetus*) of the Bering–Chukchi–Beaufort seas stock'. *Arctic*, 47: 247–55.

Craighead, J., Suydam, G. & Suydam, R. (1998) 'Observations of killer whale (*Orcinus orca*) predation in the northeastern Chukchi and western Beaufort seas', *Marine Mammal Science*, 14: 330–2.

Goley, P. D. & Straley, J. M. (1994) 'Attack on gray whales (*Eschrichtius robustus*) in Monterey Bay, California, by killer whales (*Orcinus orca*) previously identified in Glacier Bay, Alaska', *Canadian Journal of Zoology*, 72: 1528–30.

Guinet, C. (1991) 'Intentional stranding apprenticeship and social play in killer whales (*Orcinus orca*)', *Canadian Journal of Zoology*, 69: 2712–6.

Guinet, C. (1992) 'Predation behavior of killer whales (*Orcinus orca*) around Crozet Islands', *Canadian Journal of Zoology*, 70: 1656–67.

Guinet, C., Barrett-Lennard, L. G. & Loyer, B. (2000) 'Co-ordinated attack behavior and prey sharing by killer whales at Crozet Archipelago: Strategies for feeding on negatively-buoyant prey', *Marine Mammal Science*, 16: 829–34.

Hoelzel, A. R. (1991) 'Killer whale predation on marine mammals at Punta-Norte, Argentina—Food sharing, provisioning and foraging strategy', *Behavioral Ecology & Sociobiology*, 29: 197–204.

Hoelzel, A. R. (1993) 'Foraging behaviour and socal group dynamics in Puget Sound killer whales', *Animal Behaviour*, 45: 581–91.

Hoelzel A. R., Potter C. W. & Best, P. B. (1998) 'Genetic differentiation between parapatric "nearshore" and "offshore" populations of the bottlenose dolphin', *Proceedings of the Royal Society of London, Series B-Biological Sciences*, 265: 1177–83.

Lopez, J. C. & Lopez, D. (1985) 'Killer whales (*Orcinus orca*) of Patagonia and their behaviour of intentional stranding while hunting nearshore', *Journal of Mammalogy*, 66: 181–3.

Nottestad, L. & Axelsen, B. E. (1999) 'Herring schooling manoeuvres in response to killer whale attacks', *Canadian Journal of Zoology*, 77: 1540–6.

Pliny the Elder, (1938) 'Book IX: Marine animals: whales, dolphins, fish, shellfish, etc.', *Natural History*, Heineman: London, paragraphs 12–14.

Pyle, P., Schramm, M. J., Keiper, C. & Anderson, S. D. (1999) 'Predation on a white shark (*Carcharodon carcharias*) by a killer whale (*Orcinus orca*) and a possible case of competitive displacement', *Marine Mammal Science*, 15: 563–8.

Saulitis, E., Matkin, C., Barrett-Lennard, L., Heise, K. & Ellis, G. (2000) 'Foraging strategies of sympatric killer whale (*Orcinus orca*) populations in Prince William Sound, Alaska', *Marine Mammal Science*, 16: 94–109.

Scammon, C. M. (1874) *The marine mammals of the north-western coast of North America: described and illustrated, together with an account of the American whale-fishery*, J.H. Carmany: San Francisco.

Simila, T. & Ugarte, F. (1993) 'Surface and underwater observations of cooperatively feeding killer whales in northern Norway', *Canadian Journal of Zoology*, 71: 1494–9.

Slijper, E. J. (1979) *Whales*, Hutchinson & Co: London.

Sylvester, J. (1969) *The Complete Works*, George Olms Verlagsach-Handling: Hildesheim, Germany, v. I & II.

Visser, I. (1999) 'Benthic foraging on stingrays by killer whales (*Orcinus orca*) in New Zealand waters', *Marine Mammal Science*, 15: 220–7.

Visser, I. (2002) 'Kiwi killers', *Nature Australia*, 27: 46–53.

Chapter Seven—A Family Affair

Baird, R. W. & Dill, L. M. (1996) 'Ecological and social determinants of group size in transient killer whales', *Behavioral Ecology*, 7: 408–16.

Baird, R. W. & Whitehead, H. (2000) 'Social organization of mammal-eating killer whales: Group stability and dispersal patterns', *Canadian Journal of Zoology*, 78: 2096–105.

Baird, R. W., Abrams, P. A. & Dill, L. M. (1992) 'Possible indirect interactions between transient and resident killer whales—implications for the evolution of foraging specializations in the genus *Orcinus*', *Oecologia*, 89: 125–32.

Baldridge, A. (1972) 'Killer whales attack and eat a grey whale', *Journal of Mammalogy*, 53: 898–900.

Duffield, D. A., Odell, D. K., McBain, J. F. & Andrews, B. (1995) 'Killer whale (*Orcinus orca*) reproduction at Sea-World', *Zoo Biology*, 14: 417–30.

Ford, J. K. B. (1991) 'Vocal traditions among resident killer whales (*Orcinus orca*) in coastal waters of British Columbia', *Canadian Journal of Zoology*, 69: 1454–83.

Guinet, C. & Bouvier, J. (1995) 'Development of intentional stranding hunting techniques in killer whale (*Orcinus orca*) calves at Crozet Archipelago', *Canadian Journal of Zoology*, 73: 27–33.

Hoelzel, A. R., Dahlheim, M. & Stern, S. J. (1998) 'Low genetic variation among killer whales (*Orcinus orca*) in the eastern

North Pacific and genetic differentiation between foraging specialists', *Journal of Heredity*, 89: 121–8.

Ling, T. K. (1991) 'Recent sightings of killer whales *Orcinus orca* (Cetacea: Delphinidae) in South Australia', *Transactions of the Royal Society of South Australia*, 175: 95–8.

Matkin, C. O., Ellis, G., Olesiuk, P. & Saulitis, E. (1999) 'Association patterns and inferred genealogies of resident killer whales, *Orcinus orca*, in Prince William Sound, Alaska', *Fishery Bulletin*, 97: 900–19.

Silber, G. K., Newcombe, M. W. & Pérez-Cortés, M. H. (1990) 'Killer whales (*Orcinus orca*) attack and kill and Bryde's whale (*Balaenoptera edeni*)', *Canadian Journal of Zoology*, 68: 1603–6.

Whitehead, H. (1998) 'Cultural selection and genetic diversity in matrilineal whales', *Science*, 282: 1708–11.

Whitehead, H. & Glass, C. (1985) 'Orcas attack humpback whales', *Journal of Mammalogy*, 66: 183–5.

Chapter Eight—Partners in Crime

Ashford, J. R., Rubilar, P. S. & Martin, A. R. (1996) 'Interactions between cetaceans and longline fishery operations around south Georgia', *Marine Mammal Science*, 12: 452–7.

Clarke, P. A. (2001) 'The significance of whales to the Aboriginal people of southern South Australia', *Records of the South Australian Museum*, 34: 19–35.

Clutton-Brock, J. (1987) *A natural history of domesticated animals*, Cambridge University Press: Cambridge.

Egremont, P. & Rothschild, M. (1979) 'The calculating cormorants', *Biological Journal of the Linnean Society*, 12: 181–6.

Grady, D. (1982) *The Perano Whalers of Cook Strait: 1911–1964*, Reed: Wellington.

Hatt, J. (1990) 'Fine feathered fishermen', *Harpers & Queen*, May: 196–9.

Isak, H. A. & Reyer, H. U. (1989) 'Honey guides and honey gatherers: Interspecific communication in a symbiotic relationship', *Science*, 243: 1343–4.

Melville, H. (1907) *Moby Dick*, Dent: London.

Nolan, C. P., Liddle, G. M. & Elliot, J. (2000) 'Interactions between killer whales (*Orcinus orca*) and sperm whales (*Physeter macrocephalus*) with a longline fishing vessel', *Marine Mammal Science*, 16: 658–64.

Ridoux, V. (1987) 'Feeding association between seabirds and killer whales, *Orcinus orca*, around subantarctic Crozet Islands', *Canadian Journal of Zoology*, 65: 2113–5.

Yano, K. & Dahlheim, M. E. (1995) 'Behavior of killer whales *Orcinus orca* during longline fishery interactions in the southeastern Bering Sea and adjacent waters', *Fisheries Science*, 61: 584–9.

Chapter Nine—The Keeper of Souls

Howitt, A.W. (1904) *The native tribes of south-east Australia*, Macmillan: London.

MacDonald, G. F. (1996) Haida Art, University of Washington Press: Seattle.

Mansfield-Smith, G. (undated) *Hell on earth of hearth and home? A comparative study of two Australasian shore-based whaling stations, 1890–1930*, unpublished Hons thesis, ANU.

Mathews, R. H. (1904) 'Ethnological notes on the Aboriginal tribes of New South Wales and Victoria', *Journal & Proceedings of the Royal Society of New South Wales*, 39: 203–381.

Morgan, E. (1994) *The Calling of the Spirits*, Aboriginal Studies Press: Canberra.

Epilogue

Grady, D. (1982) *The Perano Whalers of Cook Strait: 1911–1964*, Reed: Wellington.

Guinet, C. (1991) 'The killer whales (*Orcinus orca*) of the Crozet Archipelago—some comparisons with other populations', *Revue D'Ecologie-La Terre et La Vie*, 46: 321–37.

Hayteas, D. L. & Duffield, D. A. (2000) 'High levels of PCB and p,p '-DDE found in the blubber of killer whales (*Orcinus orca*)', *Marine Pollution Bulletin*, 40: 558–61.

Ross, P. S., Ellis, G. M., Ikonomou, M. G., Barrett-Lennard, L. G. & Addison, R. F. (2000) 'High PCB concentrations in free-ranging Pacific killer whales, *Orcinus orca*: Effects of age, sex and dietary preference', *Marine Pollution Bulletin*, 40: 504–15.

Visser, I. N. (1999) 'Propeller scars on and known home range of two orca (*Orcinus orca*) in New Zealand waters', *New Zealand Journal of Marine & Freshwater Research*, 33: 635–42.

INDEX